BY JOHN McPHEE

The Founding Fish
Annals of the Former World
Irons in the Fire
The Ransom of Russian Art
Assembling California
Looking for a Ship
The Control of Nature
Rising from the Plains
Table of Contents
La Place de la Concorde Suisse
In Suspect Terrain
Basin and Range
Giving Good Weight
Coming into the Country
The Survival of the Bark Canoe
Pieces of the Frame
The Curve of Binding Energy
The Deltoid Pumpkin Seed
Encounters with the Archdruid
The Crofter and the Laird
Levels of the Game
A Roomful of Hovings
The Pine Barrens
Oranges
The Headmaster
A Sense of Where You Are

The John McPhee Reader
The Second John McPhee Reader

THE CENOZOIC ERA
65 MILLION YEARS

System / Period	Series / Epoch	Stage — EUROPE	Age — NORTH AMERICA
QUATERNARY	HOLOCENE		
QUATERNARY	PLEISTOCENE	TYRRHENIAN	WISCONSINAN
QUATERNARY	PLEISTOCENE	MILAZZIAN	SANGAMONIAN / ILLINOIAN
QUATERNARY	PLEISTOCENE	SICILIAN	YARMOUTHIAN
QUATERNARY	PLEISTOCENE	EMILIAN	KANSAN
QUATERNARY	PLEISTOCENE	CALABRIAN	AFTONIAN / NEBRASKAN
TERTIARY	PLIOCENE		BLANCAN
TERTIARY	PLIOCENE	PIACENZIAN	
TERTIARY	PLIOCENE	ZANCLEAN	HEMPHILLIAN
TERTIARY	MIOCENE	MESSINIAN	
TERTIARY	MIOCENE	TORTONIAN	CLARENDONIAN
TERTIARY	MIOCENE	SERRAVALLIAN / LANGHIAN	BARSTOVIAN
TERTIARY	MIOCENE	BURDIGALIAN	HEMINGFORDIAN
TERTIARY	MIOCENE	AQUITANIAN	
TERTIARY	OLIGOCENE	CHATTIAN	ARIKAREEAN / WHITNEYAN / ORELLAN
TERTIARY	OLIGOCENE	RUPELIAN	CHADRONIAN
TERTIARY	EOCENE	BARTONIAN	DUCHESNEAN / UINTAN
TERTIARY	EOCENE	LUTETIAN	BRIDGERIAN
TERTIARY	EOCENE	YPRESIAN	WASATCHIAN
TERTIARY	PALEOCENE	THANETIAN	CLARKFORKIAN
TERTIARY	PALEOCENE	MONTIAN	TIFFANIAN / TORREJONIAN
TERTIARY	PALEOCENE	DANIAN	DRAGONIAN / PUERCAN

THE MESOZOIC ERA
175 MILLION YEARS

Millions of years before the present	System / Period	Stage — EUROPE	Age — NORTH AMERICA
65	CRETACEOUS	MAASTRICHTIAN	
	CRETACEOUS	CAMPANIAN	
	CRETACEOUS	SANTONIAN	
	CRETACEOUS	CONIACIAN	GULFIAN
	CRETACEOUS	TURONIAN	
	CRETACEOUS	CENOMANIAN	
100	CRETACEOUS	ALBIAN	
	CRETACEOUS (NEOCOMIAN)	APTIAN — GARGASIAN / BEDOULIAN	
	CRETACEOUS (NEOCOMIAN)	BARREMIAN	
	CRETACEOUS (NEOCOMIAN)	HAUTERIVIAN	COMANCHEAN
	CRETACEOUS (NEOCOMIAN)	VALANGINIAN	
	CRETACEOUS (NEOCOMIAN)	BERRIASIAN	
140	JURASSIC	TITHONIAN	
	JURASSIC	KIMMERIDGIAN	
	JURASSIC	OXFORDIAN	
	JURASSIC	CALLOVIAN	
	JURASSIC	BATHONIAN	
	JURASSIC	BAJOCIAN	
	JURASSIC	AALENIAN	
	JURASSIC	LIASSIC	
205	TRIASSIC	RHAETIAN	
	TRIASSIC	NORIAN	
	TRIASSIC	CARNIAN	
	TRIASSIC	LADINIAN	
	TRIASSIC	ANISIAN	
240	TRIASSIC	SCYTHIAN	

THE PALEOZOIC ERA
330 MILLION YEARS

System / Period	Stage — EUROPE	Age — NORTH AMERICA
PERMIAN	TATARIAN	OCHOAN
PERMIAN	KAZANIAN	GUADALUPIAN
PERMIAN	KUNGURIAN	
PERMIAN	ARTINSKIAN	LEONARDIAN
PERMIAN	SAKMARIAN	WOLFCAMPIAN
PENNSYLVANIAN (CARBONIFEROUS)	STEPHANIAN	VIRGILIAN / MISSOURIAN
PENNSYLVANIAN (CARBONIFEROUS)	WESTPHALIAN	DESMOINESIAN / ATOKAN / MORROWAN
MISSISSIPPIAN (CARBONIFEROUS)	NAMURIAN	CHESTERIAN
MISSISSIPPIAN (CARBONIFEROUS)	VISEAN	MERAMECIAN / OSAGEAN
MISSISSIPPIAN (CARBONIFEROUS)	TOURNAISIAN	KINDERHOOKIAN
DEVONIAN	FAMENNIAN	CHAUTAUQUAN
DEVONIAN	FRASNIAN	SENECAN
DEVONIAN	GIVETIAN	ERIAN
DEVONIAN	COUVINIAN / EMSIAN	ONANDAGAN
DEVONIAN	SIEGENIAN	ORISKANYAN
DEVONIAN	GEDINNIAN	HELDERBERGIAN
SILURIAN		CAYUGAN
SILURIAN	LUDLOVIAN	
SILURIAN	WENLOCKIAN	NIAGARAN
SILURIAN	LLANDOVERIAN	
SILURIAN	ASHGILLIAN	MEDINAN / CINCINNATIAN
ORDOVICIAN	CARADOCIAN	TRENTONIAN
ORDOVICIAN	LLANDEILIAN	BLACKRIVERAN
ORDOVICIAN	LLANVIRNIAN	CHAZYAN
ORDOVICIAN	ARENIGIAN	CANADIAN
ORDOVICIAN	TREMADOCIAN	
CAMBRIAN	DOLGELLIAN / FESTINIOGIAN / MAENTWROGIAN	CROIXIAN
CAMBRIAN	MENEVIAN	ALBERTAN
CAMBRIAN	SOLVAN	
CAMBRIAN	CAERFAIAN	WAUCOBAN

PRECAMBRIAN TIME
ABOUT 4000 MILLION YEARS

Millions of years before present	System / Period	Regional Time Scales — EUROPE	NORTH AMERICA
570	THE PROTEROZOIC EON — HADRYNIAN	TORRIDONIAN	KEWEENAWAN
1000	THE PROTEROZOIC EON — HELIKIAN (NEO-HELIKIAN / PALEO-HELIKIAN)	LEWISIAN	
2000	THE PROTEROZOIC EON — APHEBIAN	LEWISIAN	HURONIAN
2500	THE ARCHEAN EON		LAURENTIAN
	THE ARCHEAN EON		KEEWATINIAN
4600			

Assembling California

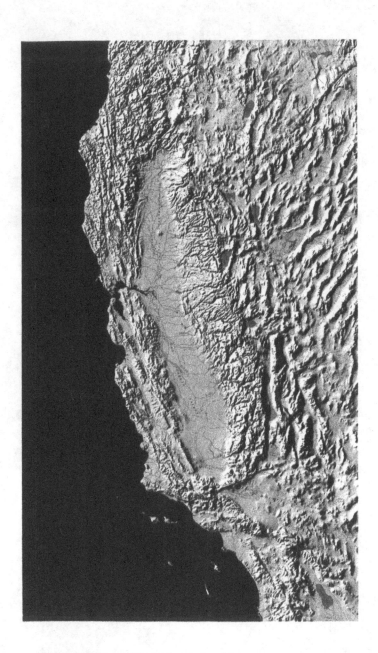

John McPhee

Assembling

California

Farrar, Straus and Giroux

New York

Farrar, Straus and Giroux
18 West 18th Street, New York 10011

Printed in the United States of America
Published in 1993 by Farrar, Straus and Giroux
First paperback edition, 1994

Nearly all of the text of this book originally
appeared in The New Yorker.

Geological time scale adapted by Tom Funk from F.W.B.
van Eysinga's Geological Time Table (*Elsevier Scientific
Publishing Company, The Netherlands*) and A Geologic
Time Scale (*W. B. Harland et al., Cambridge University
Press*) in conformity with publications of the United States
Geological Survey and the Geological Society of America.
These endpapers have been revised since the publication
of Basin and Range, In Suspect Terrain, *and* Rising from
the Plains *on the basis of recent radiometric dating.*

Frontispiece: detail from map of Landforms and Drainage
of the 48 States © 1992 by Raven Maps & Images

Library of Congress Cataloging-in-Publication Data
McPhee, John A.
Assembling California / John McPhee.
p.   cm.
1. Geology — California.   I. Title.
QE89.M37   1993   557.94 — dc20   92-35027   CIP

Paperback ISBN-13: 978-0-374-52393-0
Paperback ISBN-10: 0-374-52393-2

Designed by Cynthia Krupat

www.fsgbooks.com

26

To Kenneth Stover Deffeyes

Assembling California

Y ou go down through the Ocean View district of San Francisco to the first freeway exit after Daly City, where you describe, in effect, a hairpin turn to head north past a McDonald's to a dead end in a local dump. It is called the Daly City Scavenger Company. You leave your car and walk north on a high contour some hundreds of yards through deep grasses until a path to your left takes you down a steep slope a quarter of a mile to the ocean. You double back along the water, south to Mussel Rock.

Mussel Rock is a horse. As any geologist will tell you, a horse is a displaced rock mass that has been caught between the walls of a fault. This one appeared to have got away. It seemed to have strained successfully to jump out of the continent. Or so I thought the first time I was there. It loomed in fog. Green seas slammed against it and turned white. It was not a small rock. It was like a three-story building, standing in the

Pacific, with brown pelicans on the roof. You could walk out on a ledge and look up through the fog at the pelicans. When you looked around and faced inland, you saw that you were at the base of a fifty-foot cliff, its lithology shattered beyond identification. A huge crack split the cliff from top to bottom and ran on out through the ledge and under the waves. After a five-hundred-mile northwesterly drift through southern and central California, this was where the San Andreas Fault intersected the sea.

I went to Mussel Rock that foggy afternoon in 1978 with the geologist Kenneth Deffeyes. I have returned a number of times since, alone or in the company of others. With regard to the lithosphere, it's a good place to sit and watch the plates move. It is a moment in geography that does your thinking for you. The San Andreas Fault, of course, is not a single strand. It is something like a wire rope, as much as half a mile wide, each strand the signature of one or many earthquakes. Mussel Rock is near the outboard edge of the zone. You cannot really say that on one side of the big crack is the North American Plate and on the other side is the Pacific Plate, but it's tempting to do so. Almost automatically, you stand with one foot on each side and imagine your stride lengthening—your right foot, say, riding backward toward Mexico, your left foot in motion toward Alaska. There's some truth in such a picture, but the actual plate boundary is not so sharply defined. Not only is the San Andreas of varying width in its complexity of strands, it is merely the senior fault in a large family of more or less parallel faults in an over-all swath at least fifty miles wide. Some of the faults are to the west and under the ocean; more are inland. Whether the plate boundary is five miles wide or fifty miles wide or extends all the way to central Utah is a matter that geologists currently debate. Nonetheless, there is granite under the sea off Mussel Rock that is evidently from the southern Sierra Nevada, has travelled three hundred miles along the San Andreas system, and continues to move northwest. As evidence of the motion of the plates, that granite will do.

For an extremely large percentage of the history of the world, there was no California. That is, according to present theory. I don't mean to suggest that California was underwater and has since come up. I mean to say that of the varied terranes and physiographic provinces that we now call California nothing whatever was there. The continent ended far to the east, the continental shelf as well. Where California has come to be, there was only blue sea reaching down some miles to ocean-crustal rock, which was moving, as it does, into subduction zones to be consumed. Ocean floors with an aggregate area many times the size of the present Pacific were made at spreading centers, moved around the curve of the earth, and melted in trenches before there ever was so much as a kilogram of California. Then, a piece at a time—according to present theory—parts began to assemble. An island arc here, a piece of a continent there—a Japan at a time, a New Zealand, a Madagascar—came crunching in upon the continent and have thus far adhered. Baja is about to detach. A great deal more may go with it. Some parts of California arrived head-on, and others came sliding in on transform faults, in the manner of that Sierra granite west of the San Andreas. In 1906, the jump of the great earthquake—the throw, the offset, the maximum amount of local displacement as one plate moved with respect to the other—was something like twenty feet. The dynamics that have pieced together the whole of California have consisted of tens of thousands of earthquakes as great as that—tens of thousands of examples of what people like to singularize as "the big one"—and many millions of earthquakes of lesser magnitude. In 1914, Andrew Lawson, writing the San Francisco Folio of the Geologic Atlas of the United States, wistfully said, "Most of the faults are the expression of energies that have been long spent and are not in any sense a menace. It is, moreover, barely possible that stresses in the San Andreas fault zone have been completely and permanently relieved by the fault movement of 1906." Andrew Lawson—who named the San Andreas Fault— was a structural geologist of the first order, whose theoretical conclusions were as revered in his time as others' are at present. For the next six decades in California, a growing population tended to imagine that the stresses were indeed gone—that the greatest of historic earthquakes (in this part of the fault) had relieved the pressure and settled the risk forever. In the nineteen-sixties, though, when the work of several scientists from various parts of the world coalesced to form the theory of plate tectonics, it became apparent—at least to geologists—that those twenty feet of 1906 were a minuscule part of a shifting global geometry. The twenty-odd

lithospheric plates of which the rind of the earth consists are nearly all in continual motion; in these plate
movements, earthquakes are the incremental steps.
Fifty thousand major earthquakes will move something
about a hundred miles. After there was nothing, earthquakes brought things from far parts of the world to
fashion California.

Deffeyes and I had been working in Utah and
Nevada, in the physiographic province of the Basin and
Range. Now he was about to go east and home, and
we wandered around San Francisco while waiting for
his plane. Downtown, we walked by the Transamerica
Building, with its wide base, its high sides narrowing to
a point, and other buildings immensely tall and straight.
Deffeyes said, "There are two earthquake-resistant
structures—the pyramids and the redwoods. These
guys are working both sides of the street." The skyscrapers were new, in 1978. In an earthquake, buildings
of different height would have different sway periods,
he noted. They would "creak and groan, skin to skin."
The expansion joints in freeways attracted his eye. He
said they might open up in an earthquake, causing
roadways to fall. He called the freeways "disposable—
Kleenexes good for one blow." He made these remarks
in the shadowy space of Second Street and Stillman,
under the elevated terminus of Interstate 80, the beginnings of the San Francisco Skyway, the two-level
structure of the Embarcadero Freeway, and so many
additional looping ramps and rights-of-way that Deffeyes referred to it all as the Spaghetti Bowl. He said it
was resting on a bog that had once surrounded a tidal

creek. The multiple roadways were held in the air by large steel Ts. Deffeyes said, "It's the engineer in a game against nature. In a great earthquake, the ground will turn to gray jello. Those Ts may uproot like tomato stakes. And that will seal everyone in town. Under the landfill, the preexisting mud in the old tidal channel will liquefy. You could wiggle your feet a bit and go up to your knees." In 1906, the shaking over the old tidal channel that is now under the freeways was second in intensity only to the San Andreas fault zone itself, seven miles away. "Los Angeles, someday, will be sealed in worse than this," he continued. "In the critical hours after a great earthquake, they will be cut off from help, food, water. Take one piece out of each freeway and they're through."

In a rented pickup, we had entered California the day before, climbing the staircase of fault blocks west of Reno that had led the Donner party to the crest of the mountains named for snow. This was among the first of series of journeys on and near Interstate 80 that I would be making in the company of geologists, for the purpose of describing not only the rock exposed in roadcuts—and the regional geologies into which the roadcuts would serve as windows—but also the geologists themselves. The result was meant to be a sort of cross-section of the United States at about the fortieth parallel, and a picture of the science. The writing would develop as four compositions, of which this is the fourth. The element controlling them—the subject that has shaped the over-all structure—has been plate tectonics. The scientific papers that effected the plate-tectonics revolution were published from 1959 to 1968. Much of what was written there was at first widely scorned. As I started out on my transcontinental journeys, in 1978, I wanted to see how the science was settling down with its new theory, and, as a continuing result of its revelations, what revisions would occur in the consensual biography of the earth. Plenty of other matters would be discussed, but that one was paramount. The developed structure has not been linear—not a straightforward trip from New York to San Francisco on the interstate. It began in New Jersey and then leaped to Nevada, because the tectonics in New Jersey two hundred million years ago are being recapitulated by the tectonics in Nevada today. While the progress was not linear in a geographic sense, thematically it was aimed at California. In California was the prow of the North American Plate—in these latitudes, the sliding boundary. California was also among the freshest acquisitions of the continent. So radical and contemporary were the regional tectonics that the highest and the lowest points in the contiguous United States were within eighty miles of each other in California. As nowhere else along the fortieth parallel in North America, this was where the new theory of plate tectonics was announcing its agenda.

Over the years, I would crisscross the country many times, revisiting people and places, yet the first morning with Deffeyes among the rocks of California retains a certain burnish, because it exemplified not only how abrupt the transition can be as you move from one physiographic province to another but also the jurisdictional differences in the world of the geologist. As we crossed the state line under a clear sky and ascended toward Truckee, we passed big masses of competent, blocky, beautiful rocks bright in their quartzes and feldspars and peppered with shining black mica. The ebullient Deffeyes said, "Come into the Sierra and commune with the granite."

A bend or two later, his mood extending even to the diamond-shaped warnings at the side of the road, he said, "Falling-rock signs are always good news to us."

Then a big pink-and-buff roadcut confused him. He said he thought it consisted of "young volcanics," but preferred to let it remain "mysterious for the moment." The moment stretched. Deffeyes is as eclectic as a geologist can become, a generalist of remarkable range, but his particular expertise—he wrote his dissertation in Nevada and has done much work there since—was fading in the distance behind him. Up the road was a metasediment in dark and narrow blocks going every which way, like jackstraws. Deffeyes got out of the pickup and put his nose on the outcrop, but he had an easier time identifying a bald eagle that watched him from an overhanging pine.

"You need a new geologist," he said to me.

We took a rock sample, washed our hands in melting snow, and ate a couple of sandwiches as we watched wet traffic with bright headlights come down from Donner Summit. Looking back to the cloudless Basin and Range and seeing what lay ahead for us, Deffeyes said, "Out of the rain shadow, into the rain."

After we got up into the high country ourselves,

some additional metasediment left him colder than the rain. "The time has come to turn you over to Eldridge Moores," he said.

A few miles farther on, we came to a big, gravelly roadcut that looked like an ash fall, a mudflow, glacial till, and fresh oatmeal, imperfectly blended. "I don't know what this glop is," he said, in final capitulation. "You need a new geologist. You need a Californian."

Moores could be found on a one-acre farm in the Great Central Valley—in a tract surrounded on three sides by the vegetable-crop field labs of the University of California, Davis. Twenty years earlier, Davis had been an agricultural college, but it had since expanded in numerous directions to take its place beside Berkeley, attracting to the Geology Department, for example, such youthful figures of future reputation as the mantle petrologist Ian MacGregor and the paleobiologist Jere Lipps, not to mention the tectonicist Eldridge Moores.

At one time and another, over a span of fifteen years, Moores and I would not so much traverse California as go into it in both directions from the middle. We would hammer the outcrops of Interstate 80 from Nevada to San Francisco, reaching out to related rock even farther than Timbuctoo. Timbuctoo is in Yuba County. The better to understand California, I would follow him to analogous geological field areas in Macedonia and Cyprus—journeys much enhanced by his knowledge of modern Greek. He has read widely in Greek history as well as geologic history, and standing on the steps of the Parthenon he sounds like any other tour guide—recounting wars, explosions, orations, and stolen marbles—until he tells you where the hill itself arrived from, and when, and why the Greeks sited their temple on soluble rock that they knew to be riddled with caverns. Moores has been a counsellor through all my projects in geology, across which time our beards have turned gray. He and his wife, Judy, still live in their turn-of-the-century farmhouse, with its high ceilings, its old two-light windows, its pools of sun on cedar floors. Their children—who were five, eight, and eleven when I met them—are grown and gone. On each of two porches lie big chunks of serpentine—smooth as talc, mottled black and green. When you see rocks like that on a porch, a geologist is inside.

In the living room is a framed montage of nine covers from *Geology*, a magazine introduced in the nineteen-seventies by the Geological Society of America and raised during the editorship of Eldridge Moores (1981–88) to a level of world importance in the science. Moores is the sort of person who runs up flights of stairs circling elevator shafts, because elevators are so slow. He edited *Geology* while teaching full time and advancing his own widespread research. The montage was a gift to him from people at the G.S.A. It includes fumaroles in Iceland, dunes in southern Colorado, orange-hot lava on Kilauea, and a painting of a *Tri-*

ceratops being eaten alive by a *Tyrannosaurus rex*. In the heavens close above the struggling creatures is the Apollo Object—an asteroid, roughly six miles in diameter—that is believed to have collided with the earth and caused the extinction of the dinosaurs. In the editor's notes on the contents page, Moores referred to the painting as "the Last Supper." There were outraged complaints from geologists.

The centerpiece of the montage is a 1988 cover showing Moores on a coastal outcrop playing a cello. Moores grew up in Arizona's central highlands, in a community so remote and sparse that it was called a camp. A very great distance from pavement, it was far up the switchbacks of a mountain ridge and among the open mouths of small, hard-rock mines. At the age of thirteen, he learned to play the cello, and he practiced long in the afternoons. The miners, his father included, could not understand why he would want to do that. Moores has played with symphony orchestras in Davis and Sacramento. The coastal outcrop on the cover of *Geology* is the brecciated limestone of Petra tou Romiou, Cyprus. Moores in the field has long since overcome the most obvious drawback of a cello. He travels with an instrument handcrafted by Ernest Nussbaum in a workshop in Maryland. Essentially, it is just like any other cello but it has no belly. Neck, pegbox, fingerboard, bridge—everything from scroll to spike fits into a slim rectangular case wired to serve as an electronic belly. This is a Sherpa's cello, a Chomolungma cello, a base-camp viol. In Moores' living room is a grand piano. Still on a shelf behind it are the sheet-music boxes of his children, labelled "Brian Clarinet," "Brian Bassoon," "Kathryn Cello," "Geneva Piano," and "Geneva Violin," and three additional boxes labelled "Eldridge Cello," "Eldridge Cello and Piano," "Eldridge Cello Concertos and Trios."

Judy grew up in farming country in Orange County, New York. On her California acre of the Great Valley she grows vegetables twelve months a year, and has also raised bush strawberries, grapes, blackberries, goats, pigs, chickens, pears, nectarines, plums, cherries, peaches, apricots, asparagus, ziziphus, figs, apples, persimmons, and pineapple guavas—but not so prolifically in recent years, because she has been working with a group that provides food and emergency assistance for homeless people and for people who have run out of money and are about to be evicted. She has worked in regional science centers since she was a teen-ager, and, with others, she founded one in Davis. School buses bring children there from sixty miles around to get their hands on spotting scopes, microscopes, oscilloscopes, and living snakes, on u-build-it skeletons, on take-apart anatomies and disassembled brains. Judy, trim and teacherly, puts her hands palms down on a table to show the interaction of lithospheric plates. Lithosphere, she explains in simpler words, is crustal rock and mantle rock down to a zone in the mantle that is lubricious enough to allow the plates to move. Thumbs tucked, fingers flat, the hands side by side, she presses them hard together until they buckle upward. The hands are two continents, or other landmasses, converging, colliding—making mountains. The Himalaya was made that way. Placing the hands flat again, she slowly moves them apart. These are two plates separating, one on either side of a spreading center. The Atlantic Ocean was made that way. She begins to slide one hand under the other. This is subduction. Ocean floors are consumed that way. Thumbs tucked, fingers flat, palms again side by side, she slides one hand forward, one back, the index fingers rubbing. This is the motion of a transform fault, a strike-slip fault—the San Andreas Fault. Parts of California have slid into present place that way. Convergent margins, divergent margins, transform faults: she has outlined the boundaries of the earth's plates. There is enough complexity in tectonics to lithify the nimblest mind, but the basic model is that simple. Take your hands with you—she smiles—and you are ready for the mountains.

When I first went into the Sierra with Judy's husband, in 1978, he had an oyster-gray Volkswagen bus with a sticker on its bumper that said "Stop Continental Drift." I guess he thought that was funny. There were not a few geologists then who really would have stopped it in its tracks if they could have figured out a mechanism for doing so, but, since no one knew then (or knows for certain now) what drives the plates, no one knew how to stop them. Plate tectonics had arrived in geology just about when Moores did, and—in his metaphor—he hit the beach in the second wave. He has called it "the realization wave": when geologists began to see the full dimensions and implications of the new theory, and the research possibilities it afforded—a scientific revolution literally on a global scale. As long-established geologic concepts disintegrated under the advance of the new paradigm, people coming into geology in the nineteen-sixties realized what was being handed to them, and would carry its principles into every corner of the science, historically revising every corner of the world.

Physiographic California, for much of its length, is divided into three parts. Where Interstate 80 crosses them, from Reno to San Francisco, they make a profile that is acutely defined: the Sierra Nevada, highest mountain range in the Lower Forty-eight; the Great Central Valley, essentially at sea level and very much flatter than Iowa or Kansas; and the Coast Ranges, a marine medley, still ascending from the adjacent sea.

VERTICAL EXAGGERATION 10X

In this cross section, the Coast Ranges occupy forty miles, the valley fifty miles, the mountains ninety. All of it added together is not a great distance. It is not as much as New York to Boston. It is Harrisburg to Pittsburgh. In breadth and in profile, a comparable country lies between Genoa and Zurich—the Apennines, the Po Plain, the Alps.

An old VW bus is best off climbing the Sierra from the west. Often likened to a raised trapdoor, the Sierra has a long and planar western slope and—near the state line—a plunging escarpment facing east. The shape of the Sierra is also like an airfoil, or a woodshed, with its long sloping back and its sheer front. The nineteenth-century geologist Clarence King compared it to "a sea-wave"—a crested ocean roller about to break upon Nevada. The image of the trapdoor best serves the tectonics. Hinged somewhere beneath the Great Valley, and sharply faulted on its eastern face, the range began to rise only a very short geologic time ago—perhaps three million years, or four million years—and it is still rising, still active, continually at play with the Richter scale and occasionally driven by great earthquakes (Owens Valley, 1872). In geologic ages just before the uplift, volcanic andesite flows spread themselves over the terrain like butterscotch syrup over ice cream. Successive andesite flows filled in local landscapes and hardened flat upon them. As the trapdoor rises—as this immense crustal block, the Sierra Nevada, tilts upward—the andesite flows tilt with it, and to see them now in the roadcuts of the interstate is to see the angle of the uplift.

Bear in mind how young all this is. Until the latter part of the present geologic era, there was no Sierra Nevada—no mountain range, no rain shadow, no ten-thousand-foot wall. Big rivers ran west through the space now filled by the mountains. They crossed a plain to the ocean.

Remember about mountains: what they are made of is not what made them. With the exception of volcanoes, when mountains rise, as a result of some tectonic force, they consist of what happened to be there. If bands of phyllites and folded metasediments happen to be there, up they go as part of the mountains. If serpentinized peridotites and gold-bearing gravels happen to be there, up they go as part of the mountains. If a great granite batholith happens to be there, up it goes as part of the mountains. And while everything is going up it is being eroded as well, by water and (sometimes) ice. Cirques are cut, and U-shaped valleys, ravines, minarets. Parts tumble on one another, increasing, with each confusion, the landscape's beauty.

On the first of our numerous trips to the Sierra, Moores pulled over to the shoulder of the interstate to have a look at the outcrop that had frustrated Ken Deffeyes—the one that Deffeyes had identified as glop. It was sixteen miles west of Donner Summit, beside a bridge over the road to Yuba Gap. Moores in the field looks something like what Sigmund Freud might have looked like had Freud gone into geology. Above Moores' round face and gray-rimmed glasses and diagnostic beard is a white, broad-brimmed, canvas fedora featuring a panama block. There are weather creases at the edges of his eyes. He typically wears plaid shirts, blue twill trousers, blue running shoes. On one hip is a notebook bag, on the other a Brunton compass in a cracked leather case. He is a chunky man with a long large chest and a short stretch between his hips and the terrain. From cords around his neck dangle two Hastings Triplets, the small and powerful lenses that geologists hold close to outcrops in order to study crystals. He did not need them to see what was incorporated in this massive paradox of glop. It contained jagged rock splinters and smoothly rounded pebbles as well. "It's hard at first blush to tell that it's mudflow and not wholly glacial," Moores said. "It is mostly andesite mudflow breccia with reworked stream gravel in it and glacial till on top, which appears to be moraine but is not."

In the early Pliocene, a volcano grew into the range there. It has long since eroded away. Andesite lavas poured from the volcano. Lighter eruptive material settled around the crater. In the moist atmosphere, the volcano's eruptions caused prolonged heavy rains. The water mobilized the unstable slopes. Volcanic muds—full of the sharp rock fragments that would cement together as breccia—slid into the country. In quiet periods between eruptions, streams flowing down the volcano tumbled some of the rock fragments, rounding pebbles. In recent time, alpine glaciers dug into the country and dozed away much of what was left of the volcano, and as the ice melted it left upon the brecciated mudflows heaps of lateral till. ("It is mostly andesite mudflow breccia with reworked stream gravel in it and glacial till on top, which appears to be moraine but is not.")

All this had happened in one areal spot. All this was represented in that one roadcut. Anyone could be pardoned if, at first glance, the complete narrative seemed less than apparent. The story had repeated itself through much of the Sierra during the same band of time: other volcanoes extruding andesite and shedding mud, their remains disturbed by ice. It was a surface story, a latter-day account. The brecciated mudflows and andesite lava flows had come to rest on rock that was older by as much as five hundred million years—rock with a deep and different story, rock that just happened to be there when the mountains rose. In the discipline of stratigraphy, gaps in time are known as unconformities. The layers of the Grand Canyon are full of such temporal gaps. Much more time is absent there than is represented. If a gap of five hundred million years were the right five hundred million years, it could erase the Grand Canyon. In eastern California, the infinitesimal space between the andesite flows and the rock on which they hardened is known as the Great Sierra Nevada Unconformity. To understand what that was and how it had come to be was to understand the relationship between just two of the parts in a millipartite structure.

Moores and I went on to California's eastern boundary, turned around, and recrossed the Sierra, as we would do repeatedly in the coming years. Climbing the steep east face of the mountains, you see granite and more granite and andesite capping the granite. So far so comprehensible. But before you have crossed the range you have seen rock of such varied type, age, and provenance that time itself becomes nervous— Pliocene, Miocene, Eocene non-marine, Jurassic here, Triassic there, Ypresian, Lutetian, Tithonian, Rhaetian, Messinian, Maastrichtian, Valanginian, Kimmeridgian, upper Paleozoic. The rocks seem to change as fast as the traffic. You see olivine-rich, badly deformed metamorphic rock. You see serpentine. Gabbro. One thing follows another in a manner that seems random—a collection of relics from varied ages and many ancestral landscapes, transported from far or near, set beside or upon one another, lifted en masse in fresh young mountains and exposed in roadcuts by the state. You cannot be expected, just by looking at it, to fit it all together in mobile space and sequential time, to see in the congestion within this lithic barn—this Sierra Nevada, this atticful of objects from around the Pacific world—the events and the vistas that each item represents.

Suppose you were to find in a spacious loft a whale-oil lamp of pressed lead glass. What would you think, know, guess, and wonder about the origin and the travels of that lamp? And suppose you were to find near it a Joseph Meeks laminated-rosewood chair, and an English silver porringer and stand, and an eight-lobed dish with birds in a flowering thicket. It is possible that you would not immediately think 1850, 1833, 1662, and 1620. It is possible that you would not envision the place in which each object was made or the milieu in which it was first used, and even more possible that you would not discern how or when any of these pieces moved through the world and came to be in this loft. You also see, lined up in close ranks, a Queen Anne maple side chair, a Federal mahogany shield-back side chair, a Chippendale shell-carved walnut side chair, and a William and Mary carved and caned American armchair. Stratigraphically, they are out of order. How did that happen? Why are they here? Only one thing is indisputable: this is some loft. Jammed to the trusses, it also contains a Queen Anne carved-mahogany block-front kneehole dressing table, a Hepplewhite mahogany-and-satinwood breakfront bookcase, a rosewood Neo-Gothic chair, an Empire mahogany step-back cupboard, and a Regency mahogany metamorphic library bergère. It contains a classical brass-mounted mahogany gilt wood-and-gesso bed with pressed-brass repoussé. It contains a Federal cherry-wood-and-bird's-eye-maple bowfront chest of drawers, an early Victorian mahogany dining chair with a compressed balloon back, a Federal carved and inlaid curly-maple-and-walnut fall-front desk, a Windsor sackback writing armchair, and a Louis XV ormolu-mounted kingwood parquetry commode. There's a temple bell dating to Auspicion Day of the fifth month of the first year of Tembrun. There's a Federal carved-mahogany armchair with a cornucopian splat.

Sort that out. Complete a title search for each piece. Tell each story backward through shifting space to differing points in time. Imagine the palace, the pavilion, the house, the hall for which each piece was fashioned, the climate and location of the country outside.

Naturally, you can't do that—not in a single reconnaissance. Don't fret it. Don't fret that you can't see the story whole. You cannot tell whence each of these items has come, any more than its maker could have known where it would go.

"Nature is messy," Moores remarked. "Don't expect it to be uniform and consistent."

I remembered the sedimentologist Karen Kleinspehn saying to me in these same mountains, "You can't cope with this in an organized way, because the rocks aren't organized."

Gradually, though—outcrop to outcrop, roadcut to roadcut—Moores revived enough related scenes in the distinct origins of the random rock to frame a cohesive chronological story. That is what geologists do. "You spend a lot of time working over rocks and you have a lot of time to do nothing but think," he said. "These mountains, for example, are Tertiary normal faulted, confusing the topography with regard to structure. They show different levels of structure in different places. To see through the topography and see how the rocks lie in three dimensions beneath the topography is the hardest thing to get across to a student." After a mile of silence, he added cryptically, "Left-handed people do it better."

I said nothing for a while, and then asked him, "Are you left-handed?"

He said, "I'm ambidextrous."

As it happens, I am left-handed, but I kept it to myself.

From the east, the climb is rapid to Donner Summit—less than thirty miles, and the road is not straight. Yet elsewhere along the Sierra front the rise is so much shorter and steeper that nothing on wheels could ever climb it. From the basin below (altitude four thousand feet), you bend your neck and look ten thousand feet up a granite mass that was lifted intact, whereas here, on the route above Reno, the "Tertiary normal faulting" that Moores referred to has tiered the escarpment and lowered the crestline as well. The early trappers found a native trail here. In all likelihood, the natives who made the trail were animals, followed, in time, by people.

Under ponderosas and western cedars at the
Nevada-California line, the granite reveals itself and
then is quickly gone, as the roadside rock becomes
something like dark cordwood, fallen in columnar
blocks. This is the caprock andesite, which cracks into
columns as it cools. Another five miles, and the interstate moves through a long cut that is buff, gray, buff,
gray, and buff again as lava flows and mudflows inter-
sperse. Perhaps a hundred thousand years separate the
lava flows, while the laminating muds come ten times
as often. The volcanic cap over the granite is still a kilometre thick here. Among the trees are erratic boulders—granite boulders out of place on the andesite,
transported a few thousand years ago by a descending
ribbon of ice.

Three miles before the summit, the granite reappears, not in ice-transported bits but in bedrock at the
side of the road. And then more granite, under Jeffrey
pines—weathered granite, light and sparkling sliced
granite. It ends abruptly, at a contact with andesite.
This particular granite had been sitting here eroding
quietly for maybe ninety million years when the andesite lava flowed upon it, coating hills and filling valleys, plastering over the granitic terrain, concealing and
preserving a Miocene landscape. Differentially, ran-
domly, erosion has eaten through the caprock. So the
road encounters both formations. Granite reappears at
the summit.

Donner Summit, at seven thousand two hundred
and thirty-nine feet, is half the height of the range. Locally, engineers found a way for the interstate which is

considerably less precipitous than the trail used by the emigrants in the eighteen-forties. The place that came to be known as Donner Pass is a couple of miles south, on a relic stretch of U.S. 40. Moores and I once went over there and stood on a cliff edge, looking east. Tens of thousands of square miles of basin-and-range topography fanned out into Nevada, all of it aimed, within converging lines, at the pass. The drop to Donner Lake, more than a thousand feet below, was almost giddy. To get over the pass, everything on feet or wheels had to come up that grade. In a normal year, about seventy inches of water falls on the High Sierra, nearly all of it as snow. Seventy inches of water is roughly one and a half times what falls on New York City and twice what falls on Seattle. The snow on the Sierra Nevada can be forty feet deep. At the end of October, 1846, the Donner party came up to this pass and were forced to retreat by a mountain of snow. The winter camp where they starved and died was by the shore of Donner Lake, in the cirque below the pass.

In deep winter, I have stayed near Donner Lake in a ski condo where a previous guest left a peevish note: "The peace and beauty are marred by a noisy refrigerator and heating unit." Now, in midsummer, there were, around the pass, spreads of tenacious snow. A bicyclist, standing as he pumped but scarcely puffing, came up the route of the emigrants. Seating himself as he reached the zenith, he coasted on to the west. To the east, the deep gulf of scenery that he had come out of owed itself less to the finishing touches of ice than to large parallel north-south faults that had lowered a

large piece of country—a crustal block, dropped between two other crustal blocks, and now a graben. Lake
Tahoe, south across a partitioning ridge from Donner
Lake, lies in the same graben. The small lake and the
large one would be connected but for a recent pouring
of andesite, which formed the ridge.

Moores said to notice how the mechanical low-
ering of a large piece of the mountains had caused
varying levels of the original structure to turn up in
unexpected places. To try to sense a structure, he repeated, one must develop a talent for "seeing through
the topography" and into the rock on which the topography was carved. When rocks in their variety arrive in
a given place, like furniture going into storage, they
hold within themselves their individual histories: their
dates of solidification, their environments of deposition,
or their metamorphic experience, as the case may be.
Their unit-to-unit relationship—their stratigraphy and
other juxtapositions—pondered as a whole is structure.
Structure on the move is tectonics.

When topography is as beautiful as at Donner
Pass, it is not an easy matter to see through it, but if
you're looking for structure you might start with the
granite. In all the country from Nevada to the pass,
the volcanic cap makes its appearances, but always as
veneer—eroding everywhere, opening windows, and
ultimately suggesting the bewildering mass of the un-
derlying granite. This is the Sierra batholith. Geologists
reserve that term for the largest bodies of magmatic
rock. A batholith, as defined in the science, has a surface of at least forty square miles and no known bottom.

For the latter reason, it is also called an abyssolith. The one in California has a surface of about twenty-five thousand square miles. It lies inside the Sierra like a big zeppelin. Geologists in their field boots mapping outcrops may not have been able to find a bottom, but geophysicists can, or think they can, and they say it is six miles down. If so, the batholith weighs a quadrillion tons, and its volume is at least a hundred and fifty thousand cubic miles.

It reminds me of a big rigid airship because the rigids contained, within their metal frames, rows of giant bags that resembled aerial balloons. Batholiths develop not as single chambers of magma but as contiguous balloons of molten rock called plutons. As red-hot rising fluid, the great Sierra batholith came into the country in successive pulses during a hundred and thirty million years between early Jurassic and latest Cretaceous time. There were three peak periods—the first nearly two hundred million years before the present, the second at a hundred and forty million years before the present, and the third at eighty. The most extensive is the "80 pulse." All this went on some ten to thirty kilometres below the earth's surface, where continental crust and subducting ocean crust (coming under the continent) were melting. Through Maastrichtian time and nearly all the Cenozoic epochs, the cooled and cooling magma lay buried. The topography above changed and changed again, like a carrousel of slides. And eventually, recently, the batholith came up, to serve as the lithic medium for the erosive sculpting of Olancha Peak, of Wheeler Peak, of Mt. Whitney.

Dark cliffs above Donner Pass were Pliocene volcanics, but the rock beside the trail was granite—poeti-
cally weathered organic billows of granite. There were
small black shapes within it, like raisins. Thousands of
them. Alien pebbles. These were bits of the country
rock that the batholith intruded. They had fallen into
the magma while it was still molten or, if cooler than
that, sufficiently yielding to be receptive. They had
been softened and rounded but not melted and destroyed. On Interstate 80 west of Donner Summit, we
saw larger chips of such metasediment in the granite of
the roadcuts. Another mile, and they were larger still.
Moores referred to them as "abundant xenoliths—Jura-
Triassic pieces of the wall or the roof." These were not
the andesites and other outpourings that had been
spread upon the granite in fairly recent times; these
were parts of the intruding batholith's containing walls
or roof. They had fallen into the soft granite eighty million years ago, and, before that, had been crustal rock
for something like a hundred million years.

In the interstate median, under Jeffrey pines, were
bedrock outcrops that had been scoured and polished
by overriding ice eleven thousand years ago. There
were plenty of erratic boulders. For eleven miles after
Donner Summit, the xenoliths in the granite increased
in volume until what we had first seen as pebbles were
now the size of bears. To sense the implication of what
was coming, a structural geologist would need no further sign. We were fast approaching the wall of the
batholith—the magma's contact with the country rock.
That highway engineers would blast out a roadcut

at just such a place is fortuitous, a matter of random chance, but when Ken Deffeyes and I had come into this same right-hand bend he had shouted, "Whoa! Whoa! Pull over!" And a moment later he was saying, "This is the best outcrop on all of I-80. You can walk up and touch the wall of the great batholith."

Moores now called it "about as classic and neat a contact as you'll ever see." As cars shot past us like F-18s, he added, "Right here. Bang!" The contact was essentially vertical. It ran on up the mountainside and vanished under the trees. It could not have been more distinct had it been the line between a granite building and a brick building adjacent in a city. The granite of the batholith looked almost white beside the reddish country rock, which Moores described as the metamorphosed remains of what had once been an island arc. The granite was customary, competent—a lot of salt and less pepper. The arc rock was flaky, slaty—like aged iron in a state of ulcerated rust. In the first yards after the contact, tongues of granite reached into the country rock, preserved in the act of eating xenoliths. Within a short distance, they gave up.

As the rock ran on in the long continuous cut, it turned black, burgundy, buff, and green, in vertical stripes, in tight drapefolds with long limbs. Obviously, it had been caught up—before the arrival of the batholith—not in some minor local slumpage but in a regional and pervasive tectonic event. With two more miles, the story again made a radical change, as we came to a roadcut of gabbro. Charcoal-gray and sparkling, it was perfect gabbro. There are rich, handsome houses on the Upper East Side of Manhattan that are made of less perfect gabbro. Gabbro, too, is cooled magma. Lacking quartz, it is at the dark end of a spectrum the light end of which is, for the most part, granite. Peridotite—the rock of the earth's mantle—was in the roadcut as well, and Moores said that in his opinion these mafics and ultramafics (rocks low in silica and high in magnesium and iron) arrived after the event that had drapefolded the rock up the road and before the intrusion of the batholith. As we moved on, the gabbro-peridotite interdigitated with granite and then disappeared as the road once again descended into the Sierra batholith. After corridors of granite, there were more volcanics, in the topographic scramble of structure.

When panoramic views came along, they showed the uniformity of the sixty-mile slope—the low-angle plane of the western Sierra. The great surface (the top of the trapdoor) was completed in the eye rather than the rock. It was deeply eaten out by river gorges. To the north and the south, the vistas were wide over deep valleys to tilting planar skylines. We came to Emigrant Gap; where the erosional dissection was particularly deep. Nineteen miles from Donner Pass, the scene demonstrated with emphasis that once emigrants were across the summit they were scarcely free of trouble. From Emigrant Gap into Bear Valley they lowered their wagons on ropes. We looked into the valley, where an alpine meadow was flanked with incense cedars. Above it to the north, under the smoothly sloping skyline, were west-dipping sediments that Moores described as mudflow breccias over Paleozoic sandstones. A deep gorge cut through this ridge. It contained the Yuba River, where the Yuba, with the help of alpine ice, had captured the Bear. The two rivers, each eroding headward from opposite sides of the ridge, had struggled toward each other until the divide between them broke down, and the Bear, giving up its direction of flow, joined the Yuba and went the other way. To the northeast, under high white peaks, was a lake gouged in granite by an alpine glacier, which had left its moraines on the volcanic muds among the sharp shards and round pebbles that had caused Deffeyes to throw in his towel. Rocks between us and that lake, Moores said, were "lower Paleozoic quartz-rich sediments metamorphosed and folded at least twice." And the rocks in the peaks above the lake were remains of a Jurassic island arc.

Moores spoke reflectively of "the joy of being alone with the geology," of spending enough time walking such a scene to learn how some of it fits together, and then adding what you can to the scientific literature, "which is not like a solo but like an orchestral piece." To Moores, what had happened to create California where no surviving rock had been before was much in evidence in the scene around us, as it had been in the rocks we saw along the road. As terrane—the homonym that refers not merely to surface configurations but to a full three-dimensional piece of the earth's crust—this region had become known in geology as Sonomia. It reached from the Sonoma Range in central Nevada to the Sierra foothills west of where we stood. As plate theorists reconstruct plate motions backward through time, they see landmasses converging to form

superterranes and breaking apart to form new continents. Swept up in these great events are islands and
island arcs—Newfoundlands, Madagascars, New Zea-
lands, Sumatras, Japans—that slide in or collide in
toward continental cores. They become the outermost
laminations of new landscapes.

When terranes coming via the ocean attach themselves to a continent, they are said to have "docked."
Never shy about metaphors, geologists are not encum-
bered by the fact that they also call the docking place a
"suture." In early Triassic time, in the narrative according to present theory, Sonomia docked against western
North America. The suture is on the longitude where
Golconda, Nevada, is now. For a century or so before
plate tectonics, the obviously overriding rock was
known as the Golconda Thrust. It was an event that
happened about two hundred and fifty million years before the present. Sonomia was an island arc. North to
south, Moores said, it might have stretched two thousand miles. It brought with it those Paleozoic sandstones above Bear Valley and the quartz-rich sediments
we could also see to the northeast. Volcanoes grew in
the newly docked terrane. Bits of them would become
the xenoliths in the granites of the summit. Along the
western margin of Sonomia, where ocean crust was
subducting in a trench, more volcanoes developed.
Their rock was in peaks above us. Roughly where we
stood, a coastal region of exceptional beauty had lain at
the base of the volcanoes. Stratovolcanoes. Kilimanjaros
and Fujis.

Sonomia was actually the second terrane to attach

itself to the western edge of ancestral North America. The first had arrived in Mississippian time. It had thrust itself almost to Utah. At this latitude, a third terrane would follow Sonomia in the Mesozoic, smashing into it with crumpling, mountain-building effects that would propagate eastward through the whole of Sonomia, metamorphosing its sediments—turning siltstones into slates, sandstones into quartzites—and folding them at least twice: the multicolored drapefolds we had seen beside the road. This was the country rock the batholith intruded.

A granite batholith will not appear just anywhere. You will wait eons for one to develop under Kansas. A great tectonic event must come first. Then granite—or, rather, the magma that will cool and produce granite—comes in beneath the mountains. Volcanoes appear at the surface. Lava flows.

To create the magma, you must in some way melt the bottom of the crust. Subduction—one plate sliding beneath another—will cause things to melt. And so will a collision that compresses and thickens terrane. After a continent-to-continent collision, the crust might double; a batholith will come up within thirty million years. In deep burial, the heat from such radioactive and universal elements as uranium, potassium, and thorium is trapped. The heat increases until the rocks melt themselves and their surroundings. Granite should be forming under Tibet at present, where India has hit the Eurasian Plate in a collision that is not yet over. Under California, both thickened crust and plate-under-plate subduction contributed to the making of the batholith,

at first after Sonomia came in and sutured on and deformed itself, and again after Sonomia was hit from the
west and further deformed.

The Sierra batholith is melted crust of oceanic origin as well as continental. Most of the world's great
batholiths are not quite true granite but edge on down
the darkening spectrum and, strictly speaking, are
granodiorite. Too strictly for me. But that is the rock
of the High Sierra, which almost everyone refers to as
granite.

After the batholith came nothing during the many
millions of years of the Great Sierra Nevada Unconformity. At any rate, nothing from those years was left for
us to see. The rock record jumps from the batholith to
the andesite flows of recent time, patches of which
Moores pointed out from the lookoff at Emigrant Gap.
A few million years ago, when lands to the east of us began to stretch apart and break into blocks, producing
the province of the Basin and Range, the Sierra Nevada
was the westernmost block to rise, lifting within itself
the folds and faults of the Mesozoic dockings, the roots
of mountains that had long since disappeared. The
chronology at Emigrant Gap ends with the signatures
of glaciation on the new mountains—the bestrewn
boulders and dumped tills, the horns, the arêtes, the
deep wide U of the Bear Valley.

I remarked that geologists are like dermatologists:
they study, for the most part, the outermost two per
cent of the earth. They crawl around like fleas on the
world's tough hide, exploring every wrinkle and crease,
and try to figure out what makes the animal move.

Moores said he begged to differ. He said the whole earth is involved in plate tectonics. The earthquake slips of subducting plates could be read as deep as four hundred miles, and seismic data were now indicating that the plates' cold ocean-crustal slabs may descend all the way to the core-mantle boundary. Bumps on the core may be related to the activity of hot spots like Hawaii, Yellowstone, and Iceland. He said he wouldn't call that dermatology.

Since Moores had learned geology in the late nineteen-fifties and early nineteen-sixties, when the theory of plate tectonics was still in a formative unheralded stage, I asked him what he had been taught. How had his teachers at the California Institute of Technology explained—in what is now known as the Old Geology—the building of mountains, the rise of volcanoes, the construction of North America west of Salt Lake?

Geologists used to accept the idea that the earth's skin contracted, he said—"shrivelled up like the skin of an apple" was the favorite simile—and the wrinkles were mountains. This was said to have happened in different places at differing times, and the wrinkling process was known as an orogeny. The rise of the mountains in Utah and Nevada, where that first exotic terrane came in, was known in the old geology as the Antler Orogeny. The next wrinkling of the regional skin was the Sonoma Orogeny. The Appalachians were built in the Avalonian, Taconic, Acadian, and Alleghenian Orogenies. The Rockies were built in the Laramide Orogeny. Mountain-building mechanisms have been restyled, but these terms for them survive.

If wrinkling was the force that lifted mountain belts, it did not explain their great volume. In the second half of the nineteenth century, James Hall, the state geologist of New York, theoretically resolved that question when he conceived of what came to be called the geosynclinal cycle, and so put in place the geology that prevailed until 1968, when plate tectonics was nailed to the church door. Since mountain belts tended to rise at the margins of continents and to contain, among other things, folded marine sediments and intruding batholiths, Hall imagined a long wide seafloor trough, a deep dimple, in which vast amounts of sediment would pile up and where magmas would intrude. After a sufficient amount of material had collected, it was ready to rise, to wrinkle as mountains. Wary of the apple-skin hypothesis, many geologists preferred to think that as a geosyncline gained weight it would press down on the mantle until its volume was so great that it would rebound isostatically, like a huge buoyant log coming up from underwater. Wary of isostasy as well, many more geologists would not venture further than to say (indisputably) that "earth forces" or "orogenic forces" had lifted the geosynclines, and that these forces were "not well understood."

The geosyncline, like any admirable and serviceable fiction, contained a lot of truth. From stratigraphy to structure, geology was understood in terms of geosynclines for about a hundred years. You found gold with your knowledge of geosynclines. You found silver, antimony, and oil. You started conceptually with a geosyncline and projected events forward in time until you saw the geosyncline shuffled up in the mountains before you. Or you started with the mountains, disassembled them in your mind, and made palinspastic reconstructions, backward in time, as far as the geosyncline. The entire procedure—from the making of rock to the making of mountains to the destruction of mountains to the making of fresh formations of rock—was the geosynclinal cycle.

Inevitably, the concept was improved, refined, unsimplified. The archetypical geosyncline was deep in the middle and shallow at the sides, and grew different kinds of rocks in various places. The German tectonicist Hans Stille proposed the names miogeosyncline and eugeosyncline for the shallows and the deeps. The vocabulary was universally accepted. Miogeosynclines were the source of shallow-water sediments (limestone, for example) and no volcanics. In the eugeosynclines, volcanism occurred, and deepwater sediments, like chert, collected. In the twentieth century, as the science matured and thickened, mio- and eu- became inadequate to prefix all the differing synclinal scenes that new generations of geonovelists were describing. The germinant term was soon popping like corn. The professional conversation came to include parageosynclines, orthogeosynclines, taphrogeosynclines, leptogeosynclines, zeugogeosynclines, paraliageosynclines, and epieugeosynclines.

Moores had entered Caltech in 1955. "In the Old Geology, one learned of the eugeosyncline and miogeosyncline of western North America, which started in the late Precambrian and went through the Cretaceous," he said, in recapitulation. "Rock deformed by orogeny—folding and thrusting—from the center of the eugeosyncline out toward the continental shelf. The mechanism was 'orogenic forces.' Here in the Sierra, for example, you had a eugeosyncline and a miogeosyncline, and the eugeosyncline was thrust on the miogeosyncline. And that was the Golconda Thrust. No one knew how this 'orogeny' happened."

If California rock was disassembled on paper and palinspastically reassembled as the original geosyncline, there were shallow-water sediments followed by deep-water material, but there was no other side. "That was never explained," Moores went on. "Also, the geosynclinal cycle was said to be about two hundred million years. In the Overthrust Belt in Montana, forty thousand feet of Precambrian sediment had been thrust over Cretaceous sediment. As students, we wondered why all that Precambrian was still there. What had the source geosyncline been doing sitting there for a billion years when the cycle was two hundred million? There was no answer."

Hall's idea was not preposterous. It was incomplete. There was, after all, marine rock in mountains. Between the geosyncline and the mountains, though, something was missing, and what was missing was plate tectonics.

Wе continued west from Emigrant Gap through cuts in unsorted glacial till, buff and bouldery, and past the many blue doors of the pink garage of the Transportation Department's mountain center for snowplows and road maintenance, situated, with its cavalry of trucks, within a slowly moving earthflow, a creeping descent of unstable moraine, a sedate land-slide. "The engineers strike again," Moores said, but in scarcely three miles his contempt went into a subduc-tion zone, melted, and came back up as appreciation for a long high competent roadcut that exposed bright beds of rhyolite tuff. Twenty-nine million years ago, this air-fall ash came out of a volcano in what is now Nevada, he said, as he pulled over to the side of the road, got out, and put his nose on the engineered outcrop. While he examined the tuff through his hand lens, an eighteen-wheeler that had also come from Nevada was smoking down the mountain grade. Its brakes were furiously burning, and emitted a dark cloud. Long after the truck had gone, the cloud hung stinking in the air. The ash had been launched in several eruptive episodes. Blown west, it had landed hot, and had welded solid in successive bands. Here, more than sixty miles from the source volcano, a single ash fall was more than a metre thick. The ash had settled, of course, horizontally. Having risen with the Sierra, it was now tilting west. We descended past the four-thousand-foot contour, moving on among volcanic rocks five times older than the tuff and of more proximate origin: rock of the Sonomia Terrane altered in the heat and pressure of the assembly of California and weathered along the interstate into an abstract medley of red and orange and buff and white.

Now thirty miles west of Donner Summit, we were well into the country rock of California gold—the rock that was there when, in various ways, the gold itself arrived. The most obvious place to look for it was in fluviatile placers—the rubble of running streams. In such a setting it had been discovered. *Placer*, which is pronounced like Nasser and Vassar, was a Spanish nautical term meaning "sandbank." More commonly, it meant "pleasure." Both meanings seem relevant in the term "placer mining," for to separate free gold from loose sand is a good deal easier than to crack it out of hard rock. Some of the gold in the running streams of the western Sierra was traceable to the host formations from which it had eroded—traceable, for example, to nearby quartz veins that had grouted ancient fissures. Within two years of the discovery of gold in river gravel, gunpowder was blasting the hard-rock fissures. Into the quiet country of the low Sierra—between the elevations of one thousand and four thousand feet—gold seekers spread more rapidly than an explosion of moles. Their technology was as rampant as they were, and in its swift development anticipated the century to come. In 1848, the primary instrument for mining gold was a sheath knife. You pried yellow metal out of crevices. Within a year or two, successively, came the pan, the rocker, the long tom, and the sluice—variously invented, reinvented, and introduced.

There was also a third source of gold. It was found in dry gravel far above existing streams—on high slopes, sometimes even on ridges. The gravel lay in discontinuous pods. Geologists, with their dotted lines, would eventually connect them. In cross section, they were hull-shaped or V-shaped, and in some places the deposits were more than a mile wide. They had the colors of American bunting: they were red to the point of rutilance, and white as well, and, in their lowest places, navy blue. They were the beds of fossil rivers, and the rivers were very much larger than the largest of the living streams of the Sierra. They were Yukons, Eocene in age. Fifty million years before the present, they had come down from the east off a very high plateau to cross low country that is now California and leave their sorted bedloads on a tropical coastal plain. Forty million years later, when the Sierra Nevada rose as a block tilting westward, it lifted what was left of that coastal plain. It included the beds of the Eocene rivers, which were fated to become so celebrated that they would be known in world geology less often as "the Eocene riverbeds of California" than as, simply, "the auriferous gravels." Fore-set, bottom-set, point bar to cutbank, under the suction eddies—gold in varying assay was everywhere you looked within the auriferous gravels: ten cents a ton in the high stuff, dollars a ton somewhat lower, concentrated riches in the deep "blue lead."

To separate gold from gravel, you wash it. But you don't wash a bone-dry enlofted Yukon with the flow of little streams bearing names like Shirt Tail Creek. Mining the auriferous gravels was the technological challenge of the eighteen-fifties. The miners impounded water in the high country, then brought it to the gravels in ditches and flumes. In five years, they built five thousand miles of ditches and flumes. From a ditch about four hundred feet above the bed of a fossil river, water would come down through a hose to a nozzle, from which it emerged as a jet at a hundred and twenty miles an hour. The jet had the diameter of a dinner plate and felt as hard. If you touched the water near the nozzle, your fingers were burned. This was hydraulic artillery. Turned against gravel slopes, it brought them down. In a contemporary account, it was described as "washing down the auriferous hills of the gravel range" and mining "the dead rivers of the Sierra Nevada." A hundred and six million ounces of gold—a third of all the gold that has ever been mined in the United States—came from the Sierra Nevada. A quarter of that was flushed out by hydraulic mining.

The dry bed of an Eocene river carries Interstate 80 past Gold Run. The roadside records the abrupt change. As if you were swinging off a riverbank and dropping into the water, you go out of the metavolcanic rock and into the auriferous gravels. We stopped, stood on the shoulder, and looked about a hundred feet up an escarpment that resembled an excavated roadcut but had not been excavated by highway engineers. It was capped by a mat of forest floor, raggedly overhanging. The forest, if you could call it that, was a narrow stand of ponderosas, above an understory of manzanita with round fleshy leaves and dark-red bark. The auriferous gravels were russet, and were full of cobbles the size of tomatoes—large stones of long transport by a most impressive river.

To the south, across the highway, the scene dropped off into a deep mountain valley. The near end of the valley was three hundred feet below the trees above us. The far end of the valley was nearly twice as deep. A mile wide, this was a valley that had not been a valley when wagons first crossed the Sierra. All of it had been water-dug by high-pressure hoses. It was man-made landscape on a Biblical scale. The stand of ponderosas at the northern rim was on the level of original ground.

The interstate was on a bench more than halfway up the gravel. Above us, behind the trees, were the tracks of the Southern Pacific. In the eighteen-sixties, when the railroad (then known as the Central Pacific) was about to work its way eastward across the mountains, it secured the rights to this ground before the nozzles reached it. Moores and I made our way up to the tracks, where the view to the north was over a hosed-out valley nearly as large as the one to the south, and bordered by white hydraulic cliffs. The railroad, with the interstate clinging to its hip, ran across a septum of the old terrain, an isthmus in the excavation, an unmined causeway hundreds of feet high made of gravel and gold.

This was the country of Iowa Hill, Lowell Hill, Poverty Hill, Poker Flat, Dutch Flat, Red Dog, You Bet, Yankee Jim's, Gouge Eye, Michigan Bluff, and Humbug City. It was the country of five hundred camps that sprang up for many dozens of miles to the north and south of the present route of I-80. For a year or two, it had been a center of world news, and for some decades had clanged with industry. Now, in the dry air, nothing was stirring, not even a transcontinental freight. But looking down the two sides into the artificial valleys you could almost hear the water jets and the caving slide of gravel. Poverty Hill yielded four million dollars' worth of gold. You Bet yielded three. Humbug City got its name from a lack of confidence in the claims there, but when five million dollars came out of forty million cubic yards of flushed-away ground the name was changed to North Bloomfield. The water-dug valleys below the ground where we stood had yielded six million dollars in gold.

Yankee Jim was an Australian. A red-dog bank was a savings-and-loan ahead of its time. It issued notes in excess of its ability to redeem them. Across most of the Sierra, Interstate 80 runs close by the line of two counties—Placer and Nevada—which together produced five hundred and sixty million dollars in gold. Trans-

lated into modern values, that would be five billion.
Yankee Jim was hanged in his eponymous town.

The ancient riverbed beneath us evidently passed
through Gold Run, picked up a fossil tributary coming
in through Dutch Flat, and went off to the northwest
via Red Dog and You Bet. Before human beings appeared on earth, glacial ice and modern streams and
other geologic agents had obliterated large parts of
the Eocene river system. People had come near elim-
inating the rest. "Man is a geologic agent," Moores
said, with a glance that swept the centennial valleys.
Erosion occurs, for the most part, in what geologists
call catastrophic events—hurricanes, rockslides, raging
floods—and in that category full credentials belonged
to hydraulic mining, for scouring out and taking away
thirteen thousand million cubic yards of the Sierra.

I remember Moores rapping his geologic hammer
on an outcrop of olivine in northern Greece. He was
drawn to the rock for academic reasons, but he remarked that it might be gone before long, because of its
use in a brick that is resistant to very high temperatures. I asked him how he felt about being in a profession that calls attention to the olivine that people
tear up mountainsides to take away. He said, "Schizophrenic. I grew up in a mining family, a mining town,
and when I got out of there I had had it with mining.
Now I am a member of the Sierra Club. But you have
to face the fact that if you are going to have an industrial society you must have places that will look terrible.
Other places you set aside—to say, 'This is the way it
was.' "

I remember him referring to the same disease in response to my asking him, one day in Davis, what effect his professionally developed sense of geologic time had had upon him. He said, "It makes you schizophrenic. The two time scales—the one human and emotional, the other geologic—are so disparate. But a sense of geologic time is the most important thing to get across to the non-geologist: the slow rate of geologic processes—centimetres per year—with huge effects if continued for enough years. A million years is a small number on the geologic time scale, while human experience is truly fleeting—all human experience, from its beginning, not just one lifetime. Only occasionally do the two time scales coincide."

When they do, the effects can be as lasting as they are pronounced. The human and the geologic time scales intersect each time an earthquake is felt by people. They intersect when mining, of any kind, begins. After 1848, when the two time scales intersected in the gold zone of the western Sierra, California was populated so rapidly that it became a state without ever being a territory. As the attraction diminished, newcomers ricocheted eastward, in sunburst pattern—to Idaho, to Arizona, to Nevada, New Mexico, Montana, Wyoming, Utah, Colorado—finding zinc, lead, copper, silver, and gold, and transmogrifying the West in a manner more pervasive than the storied transition from bison to cattle. The event of 1848 in California led directly to the discovery of gold in Australia (after an Australian miner who rushed to the Sierra saw auriferous facsimiles of New South Wales). By 1865, at the end of the American Civil War, seven hundred and eighty-five million dollars had come out of the ground in California, making a difference—possibly *the* difference—in the Civil War. The early Californian John Bidwell, an emigrant of 1841, expressed this in his memoirs:

It is a question whether the United States could have stood the shock of the great rebellion of 1861 had the California gold discovery not been made. Bankers and business men of New York in 1864 did not hesitate to admit that but for the gold of California, which monthly poured its five or six millions into that financial center, the bottom would have dropped out of everything. These timely arrivals so strengthened the nerves of trade and stimulated business as to enable the government to sell its bonds at a time when its credit was its life-blood and the main reliance by which to feed, clothe, and maintain its armies. Once our bonds went down to thirty-eight cents on the dollar. California gold averted a total collapse and enabled a preserved Union to come forth from the great conflict.

Moores and I returned to the shoulder of the interstate and walked along the auriferous escarpment. The stream-rounded gravels, asparkle with quartz, are so compactly assembled there that they suggest the pebbly surface of a wide-wide screen. One does not need a director, a film, or rear projection to look into the bright stones and see the miners in motion: the four thousand who are in the region by the end of '48, the hundred and fifty thousand who follow in the years to 1884. With the obvious exception of the natives, no one is as sharply stricken by the convergences of time as Johann Augustus Sutter. He has come into a scene in which gold is unsuspected—this blue-eyed, blond and ruddy, bankrupt Swiss dry-goods merchant, with his broad-brimmed hat and his broader belly and his exceptionally creative dream. He is thirty-six years old. He envisions a wilderness fiefdom—less than a kingdom but more than a colony—with himself as a kind of duke. On a ship called Clementine, he arrives in Monterey in 1839, accompanied by ten Hawaiians and an Indian boy once owned by Kit Carson and sold to Sutter for a hundred dollars. The Mexican government, which seeks some sort of buffer between coastal California and the encroaching United States, grants him, on an incremental schedule, a hundred thousand acres of land. Sailing around San Francisco Bay, he spends a week hunting for the mouth of the Sacramento River. Soon after he finds it, there is a collection of Hawaiian grass huts on what is now Twenty-seventh Street, in a section of downtown Sacramento still called New Helvetia. He has cannons. He builds a fort, with walls three feet thick. He does not overlook a dungeon. A roof slopes in above peripheral chambers to frame a parade ground of two acres. Sutter's goal is to develop an independent agricultural economy, and he prospers. He has a gristmill. He brings in cattle and builds a tannery. He hires weavers and makes textiles. He widens his fields of grain, and draws plans for a second gristmill. He attracts many people. He issues passports.

One of the attracted people is James Wilson Marshall, of Lambertville, New Jersey, a mechanic-carpenter-wheelwright-coachmaker who is experienced

as a sawyer. Sutter sees possibilities in cutting lumber
and floating it to San Francisco. Meanwhile, he needs
boards for the new gristmill. He sends Marshall up the
American River to a small valley framed in canyons and
backed by a mountainside of sugar pines. Like many
handsome moments in western scenery, this one is
prized by the natives, who think it is theirs. A bend in
the river touches the mountains. Marshall lays a mill-
race across the bend.

He sees "blossom" in the stream gravel and remarks that he suspects the presence of metal.

A sawyer asks him, "What do you mean by 'blossom'?"

Marshall says, "Quartz."

As the sawmill nears completion, its wheel is too
low. Water is ponding around it. The best correction
is to deepen the tailrace down through the gravel to
bedrock. Yalesumni tribesmen help dig out the tailrace,
where, early in the morning of January 24, 1848, Marshall picks up small light chips that may not be stone.

Having some general knowledge of minerals, I could not
call to mind more than two which in any way resembled
this—*sulphuret of iron,* very bright and brittle; and *gold,*
bright, yet malleable; I then tried it between two rocks, and
found that it could be beaten into a different shape, but not
broken.

He sets it on glowing coals, and he boils it in lye.
The substance shows no change.

Carrying a folded cloth containing flakes the size

of lentils, Marshall journeys to New Helvetia, and insists that he and Sutter talk behind a locked door. Sutter pours aquafortis on the flakes. They are unaffected. Sutter gets out his Encyclopedia Americana and looks under G. Using an apothecary's scales, he and Marshall are soon balancing the flakes with an equal weight of silver. Now they lower the scales into water. If the flakes are gold, their specific gravity will exceed the specific gravity of the silver. Underwater, the scales tip, and Marshall's flakes go down.

Sutter at once can see the future and is dismayed by the look of it. Who will work in his sawmill if gold lies in the stream beside it? Who will complete the gristmill? What will become of his New Helvetia, his field-and-forest canton, his discrete world, his agrarian dream? He and Marshall agree to urge others to keep the discovery a local if not total secret until the mills are finished.

Coincidentally in Mexico (that is, only five days after Marshall's visit to Sutter), Nicholas P. Trist, American special agent, who has defied orders recalling him to Washington and pressed on with negotiations, successfully concludes the Treaty of Guadalupe Hidalgo. For fifteen million dollars, Mexico, defeated in battle, turns over to the United States three hundred and thirty-four million acres of land, including California.

Sutter writes a twenty-year lease with the Yalesumni for the land around the sawmill. He agrees to grind their grain for them and to pay them, in clothing and tools, a hundred and fifty dollars a year. Seeking a validation of the lease, Sutter sends an envoy to Monterey—to Colonel Richard Mason, USA, military governor of Cali-

fornia. The envoy sets on a table a number of yellow
samples. Mason calls in Lieutenant William Sherman,
West Point '40, his acting assistant adjutant general.

Mason: "What is that?"

Sherman: "Is it gold?"

Mason: "Have you ever seen native gold?"

Sherman: "In Georgia."

Sherman bites a sample. Then he asks a soldier to
bring him an axe and a hatchet. With these he beats
on another sample until—malleable, unbreakable—it is
airy and thin. Sherman learned these tests in 1844,
when he was twenty-four, on an investigative assign-
ment in Georgia having to do with a military crime.

Mason sends a message to Sutter to the effect that
the Indians, having no rights to the land, therefore have
no right to lease it.

In an April memorandum, the editor of the *California Star* says, in large letters, "HUMBUG" to the idea
that gold in any quantity lies in the Sierra. Six weeks
later, the *Star* ceases publication, because there is no
one left in the shop to print it. Thousands come
through Sutter's Mill and spread into the country. On
the American River at the discovery site, Marshall tries
to charge tithes, but the forty-eighters ignore him and
overrun his claims. They stand hip to hip like trout fishermen, crowding the stream. Like fishermen, too, they
move on, restlessly, from cavern to canyon to flat to
ravine, always imagining something big lying in the next
pool. Indians using willow baskets wash sixteen thousand dollars. People are finding nuggets the size of
eggs. "There is a chance now for every white man now

in the country to make a fortune," says a letter written to the *New York Herald* on May 27, 1848. One white man, in some likelihood Scottish, is driven insane by the gold he finds, and wanders around shouting all day, "I am rich! I am rich!" Two miners in seven days take seventeen thousand dollars from a small gully.

In June, Colonel Mason travels from Monterey to San Francisco and on to New Helvetia to see for himself what is happening in the foothills. He takes Sherman with him. They find San Francisco "almost deserted," its harbors full of abandoned ships. Ministers have abandoned their churches, teachers their students, lawyers their victims. Shops are closed. Jobs of all kinds have been left unfinished. As Mason and Sherman cross the Coast Ranges and the Great Central Valley, they see gristmills and sawmills standing idle, loose livestock grazing in fields of ripe untended grain, "houses vacant, and farms going to waste." It is as if a devastating army had traversed a wide swath on its way to the foothills from the sea.

Sutter, in the shadow of his broad-brimmed hat— his silver-headed cane tucked under his arm—warmly greets the military officers. It is scarcely their fault that two thousand hides are rotting in the vats of his abandoned tannery, that forty thousand bushels of standing wheat are disintegrating on the stem, that work has ceased on the half-finished gristmill, that the weavers have abandoned their looms, that strangers without passports—here today, gone tomorrow—have turned his fort into a boarding house and taken his horses and killed his cattle. To short-term profit but long-term disaster his canton is doomed. His dream is drifting away like so much yellow smoke. We could follow him to his destitute farm on the Feather River and on to the East, where he dies insolvent in 1880, but better to leave him on July 4, 1848, sitting at the head of his table in the storehouse of his fort, host of a party he is giving to celebrate—for the first time in California—the independence of the United States. He has fifty guests. With toasts, entertainment, oratory, beef, fowl, champagne, Sauternes, sherry, Madeira, and brandy, he presents a dinner that costs him the equivalent of sixty thousand dollars (in the foothills' suddenly inflated prices converted into modern figures). In no way has he shown resentment that rejection has met his appeal to secure his claims in a discovery of sufficient magnitude to pay for a civil war. Seated on his right is Richard Barnes Mason. Seated on his left is William Tecumseh Sherman.

By the end of 1848, a few thousand people, spread out over a hundred and fifty miles, have removed from modern stream placers ten million dollars in gold. The forty-eighters have the best of it in 1849, because the forty-niners are travelling most of that season, at the end of which fifty thousand miners are in the country. There are a hundred and twenty thousand by the end of 1855. The lone miner all but disappears. To stay abreast of the sophisticating technologies, individuals necessarily form groups. Groups are crowded out by corporations. More and more miners make less and less money, until many independents are living hand-to-mouth and their way of life is called subsistence mining. Watching companions die of disease in Central

American jungles or drown in Cape storms, they have travelled thousands of miles in pursuit of a golden goal that has now turned into "mining for beans." Always, though, there are fresh stories going east—stories that would cause almost anyone to start thinking about trying the overland route, the isthmus, the Horn.

Growlersburg is so named because of the sound of nuggets swirling in pans.

In a deep remote canyon on the east branch of the north fork of the Feather River, two Germans roll a boulder aside and under it find lump gold. Another couple of arriving miners wash four hundred ounces there in eight hours. A single pan yields fifteen hundred dollars. The ground is so rich that claims are limited to forty-eight inches square. In one week, the population grows from two to five hundred. The place is named Rich Bar.

At Goodyears Bar, on the Yuba, one wheelbarrowload of placer is worth two thousand dollars.

From hard rock above Carson Creek comes a single piece of gold weighing a hundred and twelve pounds. After black powder is packed in a nearby crack, the blast throws out a hundred and ten thousand dollars in gold.

A miner is buried in Rough and Ready. As shovels move, gold appears in his grave. Services continue while mourners stake claims. So goes the story, dust to dust.

From the auriferous gravels of Iowa Hill two men remove thirty thousand dollars in a single day.

A nugget weighing only a little less than Leland Stanford comes out of hard rock in Carson Hill. Size of a shoebox and nearly pure gold, it weighs just under two hundred pounds (troy)—the largest piece ever found in California. Carson Hill, in Calaveras County, is in the belt of the Mother Lode—an elongate swarm of gold-bearing quartz veins, running north-south for a hundred and fifty miles at about a thousand feet of altitude. There are Mother Lode quartz veins as much as fifty metres wide.

American miners come from every state, and virtually every county. Others have arrived from Mexico, India, France, Australia, Portugal, England, Scotland, Wales, Ireland, Germany, Switzerland, Russia, New Zealand, Canada, Hawaii, Peru. One bloc of several thousand is from Chile. The largest foreign group is from China. Over most other miners, the Chinese have an advantage even greater than their numbers: they don't drink. They smoke opium, certainly, but not nearly as much as the others like to think. The Chinese miners wear outsized boots and blue cotton. Their packs are light. They live on rice and dried fish. Their brothels thrive. They are the greatest gamblers in the Sierra. They make Caucasian gambling look like penny ante.

Some of the early gold camps are so deep in ravines, gulches, caverns, and canyons that the light of the winter sun never reaches the miners' tents. If you have no tent, you live in a hole in the ground. Your backpack includes a blanket roll, a pick, a shovel, a gold pan, maybe a small rocker in which to sift gravel, a coffeepot, a tobacco tin, saleratus bread, dried apples, and salt pork. You sleep beside your fire. When you get up, you "shake yourself and you are dressed." You wear a flannel shirt, probably red. You wear wool trousers, heavy leather boots, and a soft hat with a wide and flexible brim. You carry a pistol. Not everyone resembles you. There are miners in top hats, miners in panama hats, miners in sombreros, and French miners in berets, who have raised the tricolor over their claims. There are miners working in formal topcoats. There are miners in fringed buckskin, miners in brocaded vests, miners working claims in dress pumps (because their boots have worn out). There are numerous Indians, who are essentially naked. There are many black miners, all of them free. As individual prospecting gives way to gang labor, this could be a place for slaves, but in the nascent State of California slavery is forbidden. On Sundays, while you drink your tanglefoot whiskey, you can watch a dog kill a dog, a chicken kill a chicken, a man kill a man, a bull kill a bear. You can watch Shakespeare. You can visit a "public woman." The *Hydraulic Press* for October 30, 1858, says, "Nowhere do young men look so old as in California." They build white wooden churches with steeples.

In four months in Mokelumne Hill, there is a murder every week. In the absence of law, lynching is common. The camp that will be named Placerville is earlier called Hangtown. When a mob forgets to tie the hands of a condemned man and he clutches the rope above him, someone beats his hands with a pistol until he lets go. A Chinese miner wounds a white youth and is jailed. With a proffered gift of tobacco, lynchers lure the "Chinee" to his cell window, grab his head, slip a rope around his neck, and pull until he is dead. A young miner in Bear River kills an older man. A tribunal offers him death or banishment. He selects death, explaining that he is from Kentucky. In Kentucky, that would be the honorable thing to do.

Some miners' wives take in washing and make more money than their husbands do. In every gold rush from this one to the Klondike, the suppliers and service industries will gather up the dust while ninety-nine per cent of the miners go home with empty pokes. In 1853, Leland Stanford, twenty-nine years old, opens a general store in Michigan Bluff, about ten miles from Gold Run. John Studebaker makes wheelbarrows in Hangtown.

Stanford moves to Sacramento, where he sells "provisions, groceries, wines, liquors, cigars, oils & camphene, flour, grain & produce, mining implements, miners' supplies." Credit is not in Stanford's vocabulary. Miners must "come down with the dust." They come down with the dust to Mark Hopkins, a greengrocer who, sensing greater profit in picks, shovels, and pans, goes out of produce and into partnership at Collis P. Huntington, Hardware. They come down with the dust to Charles Crocker, Mining Supplies. When the engineer Theodore Judah comes down from a reconnaissance of the Sierra with the opinion that a railroad can be built across the mountains, these merchants of Sacramento have the imagination to believe him, and they form a corporation to construct the Central Pacific. The geologic time scale, rising out of the ground in the form of Cretaceous gold, has virtually conjured a transcontinental railroad.

It leaves Sacramento in 1863, and not a minute too soon, for in a sense—which is only a little fictive—it is racing the technology of mining. As the railroad advances toward Donner Summit at the rate of about twenty miles a year, the miners are doing what they can to remove the intervening landscape. Their ability to do so has been much accelerated in scarcely a dozen years. This is the evolution of technique:

Prospectors find the fossil rivers within two years of James Wilson Marshall's discovery, and soon afterward vast acreages are full of holes that seem to have been made by very large coyotes. In the early form of mining that becomes known as coyoting, you dig a deep hole through the overburden and lower yourself into it with a windlass. You hope that your mine will not become your grave. You dig through the gravel to bedrock, then drift to the side. Some coyote shafts go down a hundred feet. One goes down six hundred. When water first arrives by ditch and flume, it not only washes excavated pay dirt but is allowed to spill downslope, gullying the gravel mountainsides and washing out resident gold. This is known as ground-sluicing, gouging, booming, or "picking down the bank." Even now the terrain is beginning to reflect the fact that these visitors are not the sort who carry out what they carry in. Jack London will write in "All Gold Canyon":

Before him was the smooth slope, spangled with flowers and made sweet with their breath. Behind him was devastation. It looked like some terrible eruption breaking out on the smooth skin of the hill. His slow progress was like that of a slug, befouling beauty with a monstrous trail.

So far, the technology is not new. From high reservoirs and dug canals, the Romans ground-sluiced for gold, as did Colombian Indians before 1500, and people in the eighteenth century in the region known as the Brazils. In the words and woodcuts of *De Re Metallica* (1556) the Saxon physician Georg Bauer, whose pen name was Georgius Agricola, comprehensively presented gold metallurgy, from panning and sluicing to the use of sheepskin:

Some people wash this kind of sand in a large bowl which can easily be shaken, the bowl being suspended by two ropes from a beam in a building. The sand is thrown into it, water is poured in, then the bowl is shaken, and the muddy water is poured out and clear water is again poured in, this being done again and again. In this way, the gold particles settle in the back part of the bowl because they are heavy, and the sand in the front part because it is light. . . . Miners frequently wash ore in a small bowl to test it.

A box which has a bottom made of a plate full of holes is placed over the upper end of a sluice, which is fairly long but of moderate width. The gold material to be washed is thrown into this box, and a great quantity of water is let in. . . . In this way the Moravians, especially, wash gold ore.

The Lusitanians fix to the sides of a sluice, which is about six feet long and a foot and a half broad, many cross-strips or riffles, which project backward and are a digit apart. The washer or his wife lets the water into the head of the sluice, where he throws the sand which contains the particles of gold.

The Colchians placed the skins of animals in the pools of springs; and since many particles of gold had clung to them when they were removed, the poets invented the "golden fleece" of the Colchians.

(Translation by Herbert Clark Hoover and Lou Henry Hoover, 1950.)

California's momentous innovation in placer mining comes in 1853, after Edward E. Matteson, a ground-sluicer, is nearly killed when saturated ground slides down upon him and knocks his pick from his hand. Matteson thinks of a way to dismantle a slope from a safe distance. With his colleagues Eli Miller and A. Chabot, he attaches a sheet-brass nozzle to a rawhide hose and bombards a hill near Red Dog with a shaped hydraulic charge. That first nozzle is only three feet long and its jet at origin three-quarters of an inch in diameter. Soon the nozzles are sixteen feet long, and are called dictators, monitors, or giants. They require ever more ditches and flumes. In the words of *Hutchings' California Magazine,* "The time may come when the whole of the water from our mountain streams will be needed for mining and manufacturing purposes, and will be sold at a price within the reach of all." Where two men working a rocker can wash a cubic yard a day, two men working a mountainside with a dictator can bring down and drive through a sluice box fifteen hundred tons in twelve hours, and this is the technology that the railroad is racing to the ground at Gold Run.

Although the nozzle has the appearance of a naval cannon, it is mounted on a ball socket and is so delicately counterbalanced with a "jockey box" full of small boulders that, for all its power, it can be controlled with one hand. Every vestige of what has lain before it—forest, soil, gravel—is driven asunder, washed over, piled high, and flushed away. At a hundred and twenty-five pounds of pressure per square inch, the column of shooting water seems to subdivide into braided pulses hypnotic to the eye, and where it crashes at the end of its parabola it sounds like a storm sea hammering a beach. In one year, the North Bloomfield Gravel Company uses fifteen thousand million gallons of water. Through the big nine-inch nozzles go thirty thousand gallons a minute. Benjamin Silliman, Jr., a founding professor of the Sheffield Scientific School, at Yale, writes in 1865, "Man has, in the hydraulic process, taken command of nature's agencies, employing them for his own benefit, compelling her to surrender the treasure locked up in the auriferous gravel by the use of the same forces which she employed in distributing it!"

To get at the deepest richest gravels, which lie in the hollows of bedrock channels, the miners dig tunnels under the beds of the fossil rivers. When they reach a point directly below the blue lead, they go straight up into the auriferous gravel, where they set up their nozzles and flush out the mountain from the inside. At Port Wine Ridge, Chinese miners make a tunnel in the gravel fifteen miles long. Surface excavations meanwhile deepen. Twelve million cubic yards of gravel are washed out of Scotts Flat, forty-seven million cubic yards of gravel out of You Bet and Red Dog, a hundred and five million cubic yards out of Dutch Flat, a hundred and twenty-eight million out of Gold Run. After a visit to Gold Run and Dutch Flat in 1868, W. A. Skidmore, of San Francisco, writes, "We will soon have deserted towns and a waste of country torn up by hydraulic washings, far more cheerless in appearance than the primitive wilderness of 1848." In the middle eighteen-sixties, hydraulic miners find it profitable to get thirty-four cents' worth of gold from a cubic yard of gravel. In a five-year period in the eighteen-seventies, the North Bloomfield Gravel Company washes down three and a quarter million yards to get $94,250. Soon the company is moving twelve million parts of gravel to get one part of gold.

As the mine tailings travel in floods, they thicken streambeds and fill valleys with hundreds of feet of gravel. In their blanched whiteness, spread wide, these gravels will appear to be lithic glaciers for a length of time on the human scale that might as well be forever. In a year and a half, hydraulic mining washes enough material into the Yuba River to fill the Erie Canal. By 1878 along the Yuba alone, eighteen thousand acres of farmland are covered. Mud, sand, cobbles—Yuba tailings and Feather River tailings spew ten miles into the Great Central Valley. Tailings of the American River reach farther than that. Broad moonscapes of unvegetated stream-rounded rubble conceal the original land. Before hydraulic mining, the normal elevation of the Sacramento River in the Great Valley was sea level. As more and more hydraulic detritus comes out of the mountains, the normal elevation of the river rises seven feet. In 1880, hydraulic mining puts forty-six million cubic yards into the Sacramento and the San Joaquin. The muds keep going toward San Francisco, where, ultimately, eleven hundred and forty-six million cubic yards are added to the bays. Navigation is impaired above Carquinez Strait. The ocean is brown at the Golden Gate.

In the early eighteen-eighties, a citizens' group

called the Anti-Debris Association is formed to combat the hydraulic miners. On June 18, 1883, a dam built by the miners fails high in the mountains—apparently because it was insufficiently engineered to withstand the pressure of high explosives. Six hundred and fifty million cubic feet of water suddenly go down the Yuba, killing six people and creating a wasteland much like the miners'. On January 9, 1884, a United States circuit court bans the flushing of debris into streams and rivers. Although the future holds some hydraulic mining—with debris dams, catch basins, and the like—it is essentially over, and miners in California from this point forward will be delving into hard rock.

Edward E. Matteson—of whom the *Nevada City Transcript* said in 1860, "His labors, like the magic of Aladdin's lamp, have broken into the innermost caves of the gnomes, snatched their imprisoned treasures, and poured them, in golden showers, into the lap of civilized humanity"—spends the last days of his life at Gold Flat, near Nevada City, working as a nighttime mine watchman and a daytime bookseller. Even in the high years of his invention, he never applied for a patent. From 1848 onward, James Wilson Marshall has been literally haunted by the fact of his being the discoverer of California gold. William Tecumseh Sherman will remember him as "a half-crazy man at best," an impression that Marshall confirms across the years as he claims to consult with spirits, asking them where he might again find gold. Newcomers to California in mid-century believe that Marshall really does have some sort of divine intuition, and—to his bitter annoyance—

follow him wherever he goes. With respect to further gold strikes, nowhere is where he goes. Drinking himself to heaven, he drips tobacco juice through his beard. It stains his shirt and dungarees. Looking so, he makes a visit home. From his family's house, on Bridge Street in Lambertville, he goes up into the country toward Marshall's Corner and the farmhouse where he was born, prospecting outcrops of New Jersey diabase, hoping to discover gold. He picks up rock samples. He carries them to a sister's house and roasts them in the oven.

At the end of the twentieth century, the small farmhouse where he was born is still standing. Part fieldstone, part frame, it has long since been divided into three apartments, enveloped in a parklike shopping center called Pennytown. A boldly lettered sign on a screen door indirectly recalls Marshall's compact with Sutter. It says, "Don't Let the Cat Out."

Beside I-80, Moores inserted a knife in the auriferous gravels and pried loose a few rounded stones. He carved them to demonstrate their softness, and said, "They are practically clay. They have weathered so much they could be in Georgia."

In the nineteenth century, some of the nuggets found in the auriferous gravels were electrum—a natural pale-yellow alloy of gold and silver. Other nuggets were full of mothy cavities, where something had been eaten away—quite possibly silver. This was some of the first evidence that California enjoyed a coincidence of golds, for electrum was not characteristic of the hard-rock gold of the Sierra. The gravels had brought those nuggets from somewhere else. Rock soft enough to carve with a knife would disintegrate if it were tumbling in the bed of a stream; therefore, it had softened after it arrived. Because gold changes shape so easily, the mothy pitting of nuggets necessarily occurred after transport, too.

That the auriferous deposits were Eocene was affirmed by the fossil plants among them. The gravels themselves, with channel deposits six hundred feet deep and stones the size of basketballs, described the power of the river that brought them, the Himalayan loft of its headwaters. Fossils of subalpine Eocene vegetation have since been found in central Nevada.

"There is gold in the Carson Range, east of the Sierra, that is like the nuggets that were found in these gravels," Moores said. "The source of some California gold is probably under Nevada now."

If something as crazy-sounding as that had been said to miners in the eighteen-fifties, the miners in all likelihood would not have been surprised, for they were familiar with geologists, and geologists were not their heroes. In 1852, at Indian Bar, a miner remarked to a doctor's wife, "I maintain that science is the blindest guide that one could have on a gold-finding expedition. Those men, who judge by the appearance of the soil, and depend upon geological calculations, are invariably disappointed, while the ignorant adventurer, who digs just for the sake of digging, is almost sure to be successful." The doctor's wife, Louisa Amelia Knapp Smith Clappe, is probably the most interesting writer who was on the scene in the early days of the gold rush. Indian Bar was close by Rich Bar, where the two Germans in the deep canyon rolled a boulder and found lump gold. The doctor and his wife became resident there in 1851. She wrote letters to a sister in Amherst, Massachusetts, which have been preserved in publication under her pseudonym, Dame Shirley. At times, she may be even more purple than the interior of the Rich Bar saloon, but when she speaks of "the make-shift ways which some people fancy essential to California life," she is hitting for distance. She speaks of "red-shirted miners . . . reclining gracefully . . . in that transcendental state of intoxication, when a man is compelled to hold on to the earth for fear of falling off." She speaks of "the Irishman's famous down couch, which consisted of a single feather laid upon a rock." And she has thoughts to add about geologists:

> Wherever Geology has said that gold *must* be, there, perversely enough, it lies not; and wherever [geology] has declared that it could *not* be, there has it oftenest garnered up in miraculous profusion the yellow splendor of its virgin beauty. It is certainly very painful to a well-regulated mind to see the irreverent contempt, shown by this beautiful mineral, to the dictates of science; but what better can one expect from the "root of all evil"?

There were prospectors in the Sierra who wore over their hearts a device they called a gold magnet, explaining that in the presence of gold the magnet tingled and shocked. There were prospectors who carried forked hazelwood rods that were said to point to gold as if it were water. The miners had as much respect for them as they had for the geologists. Over their shoulders as they took off up the canyons the miners liked to say, "Gold is where you find it." As early as 1849, the *Sacramento Placer Times* remarked:

The mines of California have baffled all science, and rendered the application of philosophy entirely nugatory. Bone and sinew philosophy, with a sprinkling of good luck, can alone render success certain. We have met with many geologists and practical scientific men in the mines, and have invariably seen them beaten by unskilled men, soldiers and sailors, and the like.

All that notwithstanding, the legislature of the new State of California created in 1860 a state geological survey, and recruited the Yale-trained and already distinguished Josiah D. Whitney to be the state geologist. Nearly everybody imagined that Whitney would investigate and catalogue places in California where the earth could be turned for a profit. Instead, he gave them paleontology, historical geology, igneous petrology, stratigraphy, structure, tectonics. He gave them the minutest points of mineralogy, and he gave them the global setting. He gave them academic geology in the form that can least be turned into capital—the disciplines that lead to understanding of the history and composition of the planet. California fired him. They fired him in the modern sense that after a few years he was defunded. His name rests on the highest mountain in the Sierra Nevada.

By the erosive scenes at Iowa Hill, Poverty Hill, Forest Hill, North Bloomfield, Michigan Bluff, Gold Run, You Bet, Dutch Flat, Poker Flat, Downieville, and Smartville—the major Eocene-river deposits—Josiah Whitney was not appalled. He liked the hydraulic diggings. They flushed away the soft stuff and exposed solid rock, the better for geologists to see.

Moores cast a final glance over the man-made valley by Gold Run, and said, "It's not all that bad. Some places like this do not look bad. They are spaced out. They are not the English industrial Midlands. I like to drive cars. I like to move rapidly from place to place. There is a price we pay. If people wish to eschew all that, let them walk. When they get rid of their cars and their hi-fi sets, their credibility will rise."

He lingered long enough for a change of mood. His voice resumed at a lower and softer register. "In a couple of hundred years we are doing a good job of extracting minerals deposited over billions of years. High-grade gold deposits are just gone. Ditto copper. The U.S. has had it. There just won't be any more until we go through a few million more years of erosion, allowing the geologic processes of secondary enrichment to take place. Meanwhile, technology must extract lower- and lower-grade resources. We don't realize what we're doing."

I said I thought that we knew what we were doing and didn't give a damn.

He said, "Americans look upon water as an inexhaustible resource. It's not, if you're mining it. Arizona is mining groundwater."

Soon we were dropping toward two thousand feet, among deeply weathered walls of phyllite, in color cherry and claret—the preserved soils of the subtropics when the unrisen mountains were a coastal plain. Geologists call it lateritic soil, in homage to the Latin word for brick. All around the Sierra, between two and three thousand feet of altitude, is a band of red soil, its color deepened by rainfall that leaches out competing colors and intensifies the iron oxide. Not only phyllites but also mica schists, shales, tuffs, and sandstones in the roadcuts were red. When the road dipped far below the rooflike plane of the western Sierra Nevada, the dissected inclines around us had the appearance of red mountains covered with manzanita.

At Weimar, a little off the highway and close to the two-thousand-foot contour, was a narrow band of serpentine, the California state rock. Moores said, "Worldwide, there is an association between serpentine and gold-bearing quartz, as there is here, in the belt of the Mother Lode. Gold-quartz deposits and serpentines just go together. Where there's a hard-rock mine, serpentine will not be far away. The relationship between serpentine and quartz-vein gold is not well understood, but the miners talked about it. It was a fact of their life." On the geologic map, the serpentines showed up as strings and pods in a rich wisteria blue, like some sort of paisley print, trending north-south, signing the Mother Lode.

Also accompanying the Mother Lode was a family of major faults, confined to a zone that was scarcely fifteen miles wide but extended, both to the north and to the south of Interstate 80, more than a hundred miles. Three of the faults crossed the highway in and close to Auburn, about twenty miles below Gold Run and thirty-five above Sacramento. Auburn, once known as Rich Dry Diggings, is now the seat of Placer County. Gold was found there in a living stream less than four months after Marshall's discovery at Sutter's Mill, and

mined hard-rock ore was still being stamped to powder
at Auburn well into the twentieth century. The placer
discovery was in Auburn Ravine, which the interstate
touches as it passes through the town and under the
Southern Pacific. In 1849, Auburn was as far uphill as
you could haul things conveniently from Sacramento in
a wagon. For people and pack mules, it was the trail-
head to the burgeoning mines. Within a few years,
Auburn had become known as the Crossroads of the
Mother Lode. In masonry walls of block schist, in windows arched with sawed soapstone, something is left in
Auburn of the Roaring Fifties.

   In Auburn Ravine, a couple of hundred yards
below the railroad overpass and exactly twelve hundred feet above sea level, the interstate had been cut
through charcoal-gray rock that had very evidently been
damaged by a great deal more than human engineering.
We pulled over as soon as the shoulder was wide
enough, and walked back to have a look. We walked
past talc schists and sheared serpentines and integral
blocks of volcanic rock separated by shear zones. The
cut that had caught Moores' attention was ten feet high
and nondescript, below gray pines and trees of heaven.
It was tight to the interstate, and tandem trailers were
screaming past us. A billboard across the road said,
"Placer Savings, It's the Extras That Count." Picking
and prying at the Sierra Nevada a roadcut at a time,
Moores had crossed the mountains showing all levels of
absorption and excitement. In the presence of unusual
rock, he variously fizzes and clicks. Now, as he leaned
into this outcrop with his lens, he began to do both.

It was fine-grained diabase, in magnification asparkle with crystals—free-form, asymmetrical, improvisational plagioclase crystals bestrewn against a field of dark pyroxene. It was a much finer diabase than you would find in, say, the Palisades Sill, across the Hudson River from Manhattan. It had cooled and frozen more rapidly, but it derived from a chemically identical magma—that is to say, *essentially* identical, there being no exact copy in geology except a Xerox of your last mistake. Had this magma been extruded into air or water it would have become basalt, but—like granite, diorite, gabbro—it had chilled and formed its crystals in the absence of both. There was a signal difference, however—far beyond cooling rates or chemical composition—between this diabase and the rock of the Palisades Sill or any magma that intrudes and then hardens as a single body. To see the difference, you did not need to make a thin section—a tiny slice of rock for a microscope slide. You did not even need the hand lens. This rock had been assembled in vertical laminations like successive layers of wallboard. It had frozen not all in one piece but in continual fashion, layer after layer—a history that could be read from one lamination to the next, like bar codes indefinitely extended. Moores, ebullient, said, "We're in Fat City." Lens to eye and leaning into the outcrop, this professed and practicing agnostic said, "God, it's fantastic! God Almighty! This is a jackpot, a tremendous bit of serendipity. We've struck gold."

Given the fact that we were at twelve hundred feet in the western foothills of the Sierra Nevada and in

close proximity to serpentine and quartz, I could be for-
given if at first I took him literally. Yet all that glistered
in this outcrop was pyroxene. Gold is where you find it,
though, and for Eldridge Moores this indeed was gold.
Unlike all the other rock we had seen as we traversed
the mountains, or were likely to see in most of the aerial world, this rock in its origin was not of any continent.
It was not from slope, not from shelf, not from lake,
stream, or land. It had no genetic relationship to continental rock. Like a blue-water fish on a farmhouse
platter, it had been moved a great distance. Only a me-
teorite could have been more out of place.

Nineteenth-century geologists would have called
this rock augite porphyrite; the miners would have
called it blue diorite or slate. It was rock of the ocean
crust. Formed at spreading centers, ocean crust gradually turns cold as it travels away from the hot rift of its
beginnings toward the deep trenches where nearly a
hundred per cent of it is consumed. Down the vertical
column from salt water to mantle rock, ocean crust has
varying components, of which these laminations are the
clearest record of lateral movement. A layer at a time,
the fluid rock is driven upward in the spreading center,
solidifies, and takes its place in the long march. Most of
this happens in the mid-oceans, in the world system of
separating boundaries of plates. It also happens in the
short, isolated, and slice-like spreading centers that develop near island arcs. In all geology where rock forms
in successive layers, the layers are initially horizontal—
with this one exception. The laminations of the ocean
crust form vertically, and remain vertical as they move

to become the floors of abyssal plains and until they disappear into trenches. In Moores' words, "This is the
only situation where age progression goes sideways."

Although the rock in this outcrop had obviously
been shattered by a very great tectonic force—and although it had some extent been recrystallized as well in
the attendant heat and pressure—neither its disfigure-
ment nor its metamorphosis had masked its structure.
The laminations—known in geology as sheeted dikes—
were as narrow as ten centimetres and as wide as
eighty. By looking closely at their edges, you could all
but see the spreading center that the accumulating rock
had slowly moved away from. Layer after layer was
glassy along its right-hand edge. The magma had cooled
quickly there after touching solidified rock. The spreading center, therefore, had been to the left. After a new
lamination of magma touched hard rock and turned
marginally to glass, the rest of the lamination froze
more slowly, forming the fine crystals. Some layers had
glassy margins on both sides. They had split the weak
center of previous and still-cooling layers. In a minor
and local way, they corrupted the chronology.

When seismology first revealed the dimensions of
the ocean crust, it proved to be surprisingly thin—
about fifteen thousand feet thin—with remarkable
uniformity all over the world. The sediments upon it,
generally speaking, are not much more than a veneer.
Rock of the ocean crust—departing from spreading
centers with bilateral symmetry, ultimately disappearing in the subduction zones—is everywhere younger
than most rock of the continents. The oldest known

continental rock was discovered east of Great Bear Lake, in the Canadian Northwest Territories, in 1989, and has a uranium-lead age of 3.96 billion years. The earth itself, according to radiometrics, is six hundred million years older than that. The oldest ocean-crustal rock that has yet been found in any seafloor in the world is early-middle Jurassic—a hundred and eighty-five million years old. That is less than one-twentieth the age of the oldest continental rock and one-twenty-fifth the age of the earth itself. From spreading to sub-duction, from creation to extinction, the ocean crust completely cleans house in fewer than two hundred million years. A lithospheric plate will typically include both continental rock and ocean crust, but trenches get rid of the ocean crust while the continents stay afloat. Since rock of the sort that Moores and I were looking at does not form on continents and will not be found under a Hudson Bay, a Sea of Okhotsk, or any epiconti-nental sea, what was it doing in Auburn, California, more than five hundred miles from the nearest abyssal floor?

Moores did not have to be asked, for if he had a tectonic and petrologic specialty this was it. He had travelled the earth to see this kind of rock. Where you found it up on dry land, it proclaimed an event in the making of new country, in the mobile history of plates. It was not a signature after a fact but a precursory sign-ing in. In its transportation from the deep and its em-placement on a continent, it was not merely a clue but an absolute statement that scenery had been shifted in an operatic manner.

Toward the end of the middle Jurassic—in the high noon of dinosaurs, about a hundred and sixty-five million years ago—an island arc like the Aleutians or Japan had moved in from the western ocean and docked here. This was the third terrane at this latitude: the one that followed Sonomia and smashed into it with crumpling, mountain-building effects that propagated eastward turning soils into phyllites, sandstones into quartzites, siltstones into slates—the metamorphics we had seen up the road. In aggregate, the three terranes extended the continent by at least four hundred miles. The third one, suturing here, had doubled the width of what is now California.

The sheeted diabase that we found in Auburn—shattered so grossly in the collision—was a part of the ocean crust at the leading edge of the third terrane. As the island arc drifted eastward and the continent westward, nearly all the intervening ocean crust was consumed, but some broke off and came to rest on the continental margin, announcing the collision.

North-south, the third terrane probably came near to being a thousand miles long. What remains of it is closer to a hundred. Its width, including the part that is under the Great Valley, is about a hundred miles, too. This ten-thousand-square-mile piece of ground, named for a gold camp some twenty-five miles north of Auburn, is known in geology as the Smartville Block.

If you look at a map of the Mother Lode—its narrow band, north-south, lying under Grass Valley, Auburn, Angels Camp, French Gulch, Confidence, and Plymouth—you are, for practical purposes, looking at a map of the Smartville suture. As a geologically immediate result of the collision, the nearby rock developed the numerous high-angle faults that now appear on the geologic map along the Mother Lode. The voluminous magmas of the batholith came into the country. Water moving down through the faults would have circulated close to—or actually in—the magma, dissolving high-temperature gold compounds, and carrying them upward to precipitate the gold in fissures. Gold, which loves itself and strongly resists combination with other elements, will go into compounds at very high temperatures. The gold that is among magmatic fluids (or in the adjacent country rock) may be combined, for example, with chlorine or sulphur. The gold compounds are "modestly" soluble, and will dissolve in the water, which picks up many other elements, too, including silicon. The hot solution rises into fissures in hard crust rock, where the cooling gold breaks away from the compounds and falls out of the water as metal. Silicon precipitates, too, filling up the fissures and enveloping the gold with veins of silicon dioxide, which is quartz. In this manner, the Smartville Block, docking in the Jurassic, not only doubled the size of central California but created its Mother Lode.

If you could pull up an acre of abyssal plain anywhere in the world—lift into view a complete column of the ocean floor, from the accumulated sediments at the top to mantle rock at the base—you would find the sheeted dikes about halfway down. In contrast to the rock columns you find all over the continents—giddy with time gaps among lithologies of miscellaneous origin and age—this totem assemblage from the oceans tells a generally consistent story. At its low end is peridotite, the rock of the mantle, tectonically altered in several ways on departure from the spreading center. Above the mantle rock lie the cooled remains of the great magma chamber that released flowing red rock into the spreading center. The chamber, in cooling, tends to form strata, as developing crystals settle within it like snow—olivine, plagioclase, pyroxene snow—but above these cumulate bands it becomes essentially a massive gabbro shading upward into plagiogranite as the magmatic juices chemically differentiate themselves in ways that relate to temperature. Just above the granites are the sheeted dikes of diabase, which kept filling the rift between the diverging plates. Above the sheeted dikes, where the fluid rock actually entered the sea, the suddenly chilled extrusions are piled high, like logs outside a sawmill. Because these extrusions have convex ends that bulge smoothly and resemble pillows, they are known in geology as pillow lavas. Above the pillows are the various sediments that have drifted downward through the deep sea: umbers, ochres, cherts, chalk. Unlike the rest of the crust-and-mantle package, the sediments may hint at the surrounding world. Water that gets down through all this and into the mantle rock—at the spreading center or anywhere else—will change the nature and appearance of that rock. Through an alteration of minerals, the rock takes on a silky lustre and a very smooth texture, becomes fibrous, and develops color—occasional streaks and spots of white, but mainly chrome green, myrtle green, Nile green, in patterned shapes within the mantle black. Because the patterns strongly suggest the skin of a snake, this rock has been known—for nearly six hundred years in the English language—as serpentine. Geologists—in their strange, synecdochical way—have named the entire oceanic assemblage for this one component rock. But not directly. In their acute sense of time, they were not content to settle for a term of Latin derivation. Instead, they extracted from a deeper stratum ὄφις— *ophis*—the Greek word for snake. From the mantle upward, the complete column of ocean-floor rock is collectively known in geology as an ophiolite. The generally consistent differences within it are the ophiolitic sequence.

On the American River under the bluffs of Auburn, in 1852, a single pan of gravel might be worth a hundred dollars. In 1857, after the lone miners had worked the place over, the American River Ditch Company built a dam there, to impound water for hydraulic mining. The dam eventually crumbled. The dam site did not. As environmentalists have discovered to their eternal chagrin, a dam site is a dam site forever, no matter what the state or the nation may decide to do about it in any given era. On present road maps of California, that part of the American River is marked "Auburn Dam and Reservoir (Under Construction)."

The dam site is scarcely a mile from the shattered ophiolite of Interstate 80, so Moores and I went to see how the dam might relate to the Smartville collision, and we have returned there since. The river's deep canyon is walled with sheared foliated rock—broken, disrupted, deformed lithologies, *Bruchgeitrochen*, tortured rocks—as one would expect of a place where an oceanic island arc had sutured onto the continent. There were sheeted dikes, serpentines, plagiogranites, gabbros, and other items from the ocean suite. The type of dam chosen in 1967 by the Department of the Interior's Bureau of Reclamation was a thin arch of concrete rising six hundred and eighty-five feet from channel to crest. Its purpose was to store winter runoff for use in summer, supplementing the storage behind Folsom Dam, fifteen miles downstream. The new reservoir, Lake Auburn, would reach twenty miles into the Sierra, filling two forks of the river—up the North Fork past Codfish Creek and Shirt Tail Creek beyond Yankee Jim's almost to Iowa Hill, and up the Middle Fork over New York Bar and Murderers Bar and the Ruck-a-Chucky Rapids to Volcanoville. The lake would cover ten thousand acres and be twice as deep as the Yellow Sea.

When Moores and I first visited the site, in 1978, it resembled one of the huge excavations flushed out by the hoses of hydraulic mining. Benched roadways descended switchbacks a thousand feet down the canyon walls. A cofferdam had tucked the river to one side. Reaching eleven hundred and fifty feet across the canyon floor lay the white concrete of the dam's base. From the outset, the construction project had had to deal with the inconvenience of the faulting that had followed the arrival of the Smartville Block. The dam site was squarely in the suture zone. Under the dam's foundation ran a fracture known to engineers as the

F-1 Fault. A tectonic event on the scale of an arc-to-continent docking will not result in every fissure's being filled with quartz and gold. Countless empty cracks remain. In order to secure the dam's basement, the Reclamation engineers had performed what they described as "dental work," a "root canal." They had sealed in the Smartville fault zone with three hundred and thirty thousand cubic yards of grout.

Moores remarked, "If you want to find a fault in California, look for a dam."

Scarcely had the dental work been completed when, in 1975, an earthquake struck near Oroville, forty-five miles up the Smartville Block. Its Richter magnitude was 5.7. Near Oroville on the Feather River was the eighth-largest dam in the world. It had been completed in 1968, only seven years before the earthquake, and gradually its reservoir had impounded forty-six hundred million tons of water, or enough to put a lot of pressure on the rocks below. It was a gravity dam, broad and squat, an earth-fill dam, and what it had been filled with was seventy-eight million cubic yards of hydraulic-gold-mine tailings. Absorbing the earthquake, the dam just sat there, holding its lake. However, the United States Geological Survey (a sibling agency of Reclamation within the Department of the Interior) quietly noted that twenty-five per cent of reservoirs of comparable depth had, by their sheer weight, triggered earthquakes. The earthquake at Oroville had been five times larger than the maximum earthquake that the dam at Auburn was designed to withstand.

This collection of facts soon assembled itself in the editorial offices of the *Sacramento Bee.* On its front page the *Bee* envisioned an earthquake that would knock out Auburn Dam, releasing water that would in less than two hours stand in Sacramento twenty feet deep. In the words of a former Assistant Secretary of the Interior, this would be "the worst peacetime disaster in American history." Estimates were that a quarter of a million people would drown.

The federal government considers faults inactive if they haven't jumped for a hundred thousand years. The latest known movement along the F-1 Fault in Auburn was in the Jurassic—a hundred and forty million years before the present—but that did not soothe Sacramento. Work on Auburn Dam was suspended. In 1978, Moores and I found the site silent, dry, reliquary. It looked that way many years later, when we went there again. Geologic time and human time seemed to have met and parted.

Mountain lions go through the dam site. Bears. Feral goats. The project is dormant and appears dead, or vice versa, depending on how the beholder eyes it. The bureau keeps a skeleton crew there, each of whom speaks of the dam in the future positive.

You ask at what altitude the lake level was to be.

Response: "The lake level will be eleven hundred and thirty feet."

You ask if the boat ramp would have been paved.

Response: "Yes, that will be paved."

The gravel boat ramp, several hundred yards long, descends a steep slope and ends high and nowhere, a dangling cul-de-sac. The skeletons call it "the largest and highest unused boat ramp in California." Houses that cling to the canyon sides look into the empty pit. They were built around the future lakeshore under the promise of rising water. You can almost see their boat docks projecting into the air. Thirty-three hundred quarter-acre lots were platted in a subdivision called Auburn Lake Trails.

Moores wanted to know if a geology student from Davis might be permitted to study the rocks that had been exposed during construction.

"Fine, but we would not want the student's conclusions to inconvenience the dam."

The dam has cost several hundred million dollars so far. The bureau spends a million dollars a year maintaining the site while nothing happens.

We looked for a cup of coffee in Cool, California, after crossing the American River on a seven-hundred-foot-high bridge. Not particularly long, the bridge was built so high in order to clear the lake that wasn't there. From houses in Cool, picture windows framed the lake air. There were numerous for-sale signs. Mother Lode Realty. Cool was a placer-mining camp of the eighteen-fifties. In Cool Quarry, marine limestone is mined now—a lenticular pod, a third of a cubic mile, shoved into California by the arriving Smartville Block. If you had lived on the moon then, as a full earth came into the sky you would have seen two large continents (Laurasia and Gondwanaland), one above the other, surrounded and divided by ocean. West to east, the dividing seas were the incipient Caribbean (Central

America was not there), the incipient Atlantic, and—
from Gibraltar through China—the long water known
in geology as Tethys. Worldwide, fossils from that time
are described as Tethyan. Tethys, mother of rivers, was
the consort of Oceanus. Cool was named for Aaron
Cool. In the limestone pod in Cool, California, caught
up in the docking of the Smartville Block, are Tethyan
fusilinids and Tethyan corals.

As an overriding plate scrapes the plate below, it
acts like the blade of a bulldozer and piles up sand,
seashells, cherts, phyllites—whatever happens to be
there. Impressive amounts of material can be accreted
in this manner. As the Philippine Plate has scraped westward, overlapping the Eurasian Plate, an accretionary
wedge has risen as Taiwan. In an arc-to-continent collision that is reducing the distance between Taipei and
Beijing, Taiwan is the first piece of the West Luzon Island Arc to reach the Eurasian slope. In the mélange
of rocks in the Taiwan accretionary wedge are not
only sands, seashells, cherts, and phyllites but enough
scraped-up ocean-floor debris—sheeted dikes, pillow
lavas, gabbros, serpentines—to be known as the East
Taiwan Ophiolite. A large and intact package of ocean
crust-and-mantle can be expected to follow, as one did
when Smartville began to close with North America.

The Smartville Block pushed before it not only
the limestone of Cool and the schists and serpentines
of Auburn Dam but also the red-weathered phyllites
and the argillites and cherts we had seen along the
interstate as we descended toward Auburn. These and
a great miscellany of additional rocks were Smart-

ville's mélange, its accretionary prism—highly foliated, sheared, broken, disrupted, deformed—caught up in the Smartville suture, the docking of arc and continent.

There was, of course, a subduction zone—a trench—between the arc and the continent as they drew together, and in the collision it disappeared. It was actually stuffed shut, according to present theory. First, ocean crust-and-mantle of the North American Plate went down the trench. Eventually, the continental rock itself reached the trench and jammed it, like a bagel in a toaster. Continents are too light and thick to be subducted, and where they arrive at trenches the trenches cease to function. Australia has jammed the trench to the north of it with such force that it has produced New Guinea. As Moores envisions the Jurassic event in California, a large overlap of Smartville ocean crust (the upper plate) was left lying on the North American continental slope after subduction stopped. The region cooled in the postcollisional stillness. The lower, west-moving plate was no longer descending, dragging everything downward. Isostasy, the force that lifts light objects when other forces cease to hold them down, began to work on the combined terranes. Lifting them, it broke off a large piece of the Smartville ocean crust-and-mantle and carried it into the air in what would eventually become the foothills of the Sierra Nevada.

As a geological and geophysical specialty, the study of ophiolites is only a few years old, and therefore provokes argument on almost any question raised—from the environment of the origin of the rock itself to the putative method by which it made its way from extracontinental depths to dry land. The emplacement story for the Smartville Block was worked out by Eldridge Moores.

Descending westward, just below Auburn, you cross the thousand-foot contour, and the Great Central Valley comes into view, running flat out of sight to the horizon. Sacramento is down there, and, fifteen miles farther, Davis. It is an abrupt, absolute change of physiographic worlds, where the mountains hinge. The Smartville Block extends under the valley and ends beneath the Coast Ranges.

After the trench at Auburn disappeared, another one had to develop, as the trailing edge of Smartville became the new front of the North American Plate, moving west. To balance the earth's books by consuming an amount of ocean crust comparable to what is made at spreading centers, subduction zones develop when and where they are needed. Across geologic time, subduction zones have come and gone quite often. On one side of the Smartville Block the new subduction must have been developing even while the old subduction died on the other. "For sure," Moores said. "You've got to produce that convergence somewhere. You don't cut things off. The subduction here at Auburn was Permian to early Jurassic. The new subduction zone to the west was operating by the late Jurassic, and it operated all through Cretaceous and into Tertiary time. The volcanoes came up in the Sierra, and the main batholith formed. I think the geology is really neat. The new plate margins produced their own accretionary prism, piling up out there to the west of us to become the rest of California."

Moores had become aware of the unusual rocks in the Sierra foothills only four years before I met him. On a spring day in 1974, he and his family left Davis for a camping trip in the mountains, and instead of following their usual route, up the Feather River, they took a right at Yuba City and went into the canyon of the Yuba. The theory of plate tectonics was six years old. Eight more years would pass before the term "exotic terrane" would be coined. Geologists, caught up in the fresh intensity of a scientific revolution, were still at the beginnings of seeing the world anew. Few people then envisioned the western United States as a collection of lithospheric driftwood. Moores, however, had worked for some years in Macedonia, struggling to understand the large unrooted masses of mantle rock that lay there on the surface as mountains, and to relate it to the stratiform gabbros, plagiogranites, and sheeted dikes nearby. With Fred Vine, now of the University of

East Anglia, he had also worked in the Troodos Mountains of Cyprus, where sheeted diabase ran on for
seventy miles, and where the gabbros, the granites,
and capping marine sediments were present as well.
Moores and Vine decided that the whole assemblage
had somehow been removed from a blue-ocean setting
and emplaced upon the African slope. In 1971, they
published in the *Philisophical Transactions of the Royal
Society* (*London*) an establishing paper, the significance
of which was expressed in its title: "The Troodos Massif,
Cyprus, and Other Ophiolites as Oceanic Crust."

Between the Great Valley and the High Sierra, the
Yuba drainage lay in the geographic center of what was
not yet known as the Smartville Block. Moores wasn't
there to do geology. All he was trying to do was to go
uphill out of the oak woodland and into the ponderosas,
a task to which his micropowered microbus was almost
unequal. It chugged, and—among the brown grasses of
a dry country—the rock at the roadside passed slowly.
At work or play, a geologist always drives like Egyptian
painting—eyes to the side. Near the tributary Dry
Creek, about ten miles north of the Yuba, the microbus
ascended a particularly formidable incline between
high roadcut walls of a dark and igneous rock. Moores
did not stop, but the hair on his arm may have moved.
He remembers feeling that he could have been in
Macedonia. More emphatically, in Cyprus.

On the geologic map of California, the lithology
of that area was vaguely described as "Jura-Triassic
metavolcanic rocks." The eye sees what it is trained
to see. Or, according to a maxim that Moores often

quotes, "the eye seldom sees what the mind does not anticipate." When Moores returned to that roadcut with a lens in his hand, he did not need it. He found— one standing beside the next beside the next beside the next—classic sheets of sparkling diabase, each with a glassy margin, all in prime condition, undeformed.

The sheeted dikes that he would later find on Interstate 80 at Auburn were deformed in the Smartville suture almost beyond recognition. These in the Yuba Valley were well back of Smartville's impacted leading edge. When he and I stopped there once, he said, "You could look at hand specimens of rocks taken from this cut and not be able to say if they were from Cyprus, Pakistan, Oman, New Guinea, Newfoundland, or California. Only by dating and by detailed chemical analysis of minor elements can you tell them apart." From Auburn Dam north to Oroville, there is a continual exposure of forty miles of these sequential laminations of seafloor spreading.

I asked him if he could say from what distance and what western compass point the Smartville Block had come.

"No," he said. "We have tried doing paleomagnetic work, but so far it's inconclusive. The Smartville are probably developed not more than a thousand kilometres from North America. It probably came from the northwest."

In altitude, the Smartville Block rises roughly from zero to five thousand feet. To move about on its small roads is to move among zones of time that are not all written in rock. U-2s drop out of the stratosphere and

hang-glide into Beale Air Force Base while you walk up
a dry streambed at Oregon House, where placers were
first panned in 1850, in a valley of sheeted diabase that
was injected into the floor of the North Pacific Ocean
about a hundred and sixty million years ago and emplaced on California five million years after that. The
ophiolite—the whole vast assemblage of transported
deep-ocean rock—now rests on California like a ship
stuck in sand, listing thirty degrees to the west. The
ophiolite tilts more steeply than the slope of the Sierra.
Therefore, as you climb the modern foothills, geologically you go downsection, ever deeper into the former
seafloor, from the spreading rift to the granites and gabbros of the magma chamber that fed it, and on to the
cumulate layers of heavy crystals that settled on the
mantle at the boundary that is known in the science as
the Moho. On up the mountains (and farther downsection) are scattered serpentines derived from the mantle
itself. Where the rock has not been folded, the descent
of the ophiolite is as clear as it is complete. It is the per-
verseness not of the science but of the earth itself that
you often go downsection when you are going uphill.

In the other direction, the ocean-crustal rock that
lay above the sheeted dikes is in the foothills also. Below Timbuctoo Bend on the Yuba, for example, a green
ledge of what appear to be massed satin pillows reaches
into the fast clear river. Notwithstanding the boulder
fields of hydraulic-mining tailings which cover the
floodplain and extend out of sight downstream, it is a
place of such appeal that you reflexively reach for a fly
rod, or look around for a place to pitch a tent. Whereas

the pillows at Auburn are so extensively crushed that only a specialist can reassemble them in the eye, the pillows of Timbuctoo are rotated but undamaged. Each about two feet in diameter, they are simple in form, elegant, ovate—a spread of huge caviar. Nowhere on land, Moores said, will you see pillow lavas more perfect than these.

Timbuctoo was a placer camp in 1849. When Moores and I first stopped there, in the nineteen-seventies, we read these words on a wall of a roofless masonry building:

GOLD DUST BOUGHT
WELLS FARGO & BROTHERS

And, barely legible in fading paint, "Stewart Brothers have for sale dry-goods, boots and shoes, ready-made clothing, groceries and provisions." That structure was all there was of Timbuctoo, where twelve hundred people lived in the eighteen-fifties. Now there is even less. The Wells Fargo building has collapsed. One masonry corner juts chimneylike above the rubble. The words are gone, and only a hill beside the river retains an indelible scar, torn out by hydraulic mining as if by rapid landslide.

In or close to the magma chambers under oceanic spreading centers, seawater, which has descended through fissures, dissolves metals (copper, silver, iron, magnesium, gold); it carries the metals upward and precipitates them on top of the new rock. If the new rock, in its migration, happens to end up on a continent, the metal comes with it. In the suturing process, faults form. Deeply circulating groundwater redissolves the metals and redeposits them in quartz veins in the faults. In this way, Moores said, it made sense to imagine that the Mother Lode gold of California came in from the deep ocean riding the Smartville Block.

Smartville, California, is a living town, 95977. It is just uphill from Sucker Flat and a mile from Timbuctoo. Sucker Flat was named for Illinois. The miners looked upon Illinois as "the Sucker State." Jim Smart, of Smartville, was not from Illinois, and he was smart enough not to be a miner. He ran a hotel. Never mind that the Sucker Flat Channel, between Timbuctoo and Smartville, yielded two and a half million dollars in gold. There is a white wooden church in Smartville, its paint peeling. There are two gas pumps under a sign that says "Bait." There are slopes of brown grass under blue oaks—and the houses of a hundred and fifty people. Roadcuts in Smartville are full of bulging pillows.

In the roadcuts of Rough and Ready, nine miles from Smartville, we saw massive gabbros. Rough and Ready was founded by forty-niners, and its citizens soon voted to secede from the Union. If geologists of the nineteenth century and the first three-quarters of the twentieth century failed to see the intrinsic bond between the gabbros of Rough and Ready and the pillow basalts of Smartville and Timbuctoo (and the relationship of those rocks to the diabase and plagiogranite and serpentine nearby), Moores was sympathetic. He said, "If you found a headlight, a hubcap, a brake drum, and a radiator, you would say, 'Ah, pieces of an automobile.' But if you had never seen those things assembled you would not relate them to each other. The ophiolitic sequence is one of the most classic things of importance to the plate-tectonics story. Its emplacement is evidence of the spreading process and of the subduction process, not to mention the consumption of vast amounts of ocean crust. There are ten thousand square miles of ground here with no one arguing about the fact that it's island-arc material. The original arc may or may not have been the size of Japan, or the Philippines, or the Marianas, or the Antilles, or the Aleutians. But it was surely an island arc, and its arrival is signalled by these ophiolitic rocks. If ophiolites are found at the suture of the Urals—as they are—it means that there was a sea between Siberia and Europe. The sea was consumed. Ophiolites were emplaced. And the Urals are welded in a Permo-Triassic suture. Before the Permo-Triassic, in the sea and the islands, the Gulag Archipelago was real."

Moores was doing postdoctoral work at Princeton University, in the middle nineteen-sixties, when he first heard about what was described to him as "a wonderful complex on Cyprus," but when he tried to look it up he could find no data. He requested, from Nicosia, memoirs of the Cyprus Geological Survey. This was at the time when the theory of plate tectonics was in its coalescing phase. The term itself did not yet exist. Oceanic spreading centers were known, and the consumption of ocean crust in subduction zones was beginning to be understood. The idea of a suite of rock called an ophiolite had been around in the science for many years but had not yet been widely accepted or related to the new theory. As the story of plate tectonics further unfolded, the story of the ophiolitic sequence would accompany it like an echo.

Fred Vine, who was also working at Princeton at the time, had co-authored with his Cambridge University colleague Drummond Matthews a paper that contributed the manifesting insight into the movement of the ocean crust and placed Vine and Matthews among the handful of people in various parts of the world who collectively brought about the plate-tectonics revolution. Vine and Matthews compared data collected by seaborne magnetometers with the known history of reversals of the magnetic field of the earth. In the last hundred million years, the magnetic pole has jumped from north to south and back again nearly two hundred times. It has remained in the one place or the other for highly varying lengths of time—as much as twenty million years and as little as twenty thousand. Derived from paleomagnetic evidence taken from the rock of continents, this peculiar time scale was aligned by Vine and Matthews with zones of strong and weak magnetism that could be mapped on seafloors using the data collected by the magnetometers. Expressed as units of time, the polar reversals and the zones of seafloor magnetism matched exactly. Under the ocean, evidently, time had two arrows, whose points touched at the mid-ocean ridge. With bilateral symmetry, the seafloor grew older with distance from its spreading center.

When Moores' packet arrived from Cyprus, he already had behind him three years' experience in

Macedonia among the lower components of ophiolitic rock, but he had not sensed their origins. He had not imagined that they had formed in one milieu and been transported to another. Like most geologists, he thought of them as rocks formed from magmas that had welled up under Greece. Opening a geologic map of Cyprus, he saw diabase dikes all running in the same direction; he saw mantle-derived serpentine at the bottom of the section and basaltic pillows at the top. This time, his imagination made the jump. Unfolding the map in front of Fred Vine, he said, "How does this look for oceanic crust formed at a spreading center?"

In the science, the ancestral glimmerings of that intuition were nearly a century old. By the eighteen-eighties, geologists had begun to reflect on a common association of serpentine, gabbro, diabase, and basalt; and in *The Face of the Earth,* Eduard Suess, of Vienna, observed that all these "green rocks," as he called them, could characteristically be found in folded-and-faulted mountains. He said that they had formed in the geosynclines from which the mountains derived.

In 1892, when the German geologist Gustav Steinmann visited San Francisco, Berkeley's Andrew Lawson took him to the north side of the bridgeless Golden Gate to see the rocks of Marin. (Moores came upon this remarkable account while he was reading Steinmann in German in 1967.) Steinmann had wandered the Apennines and Alps noticing serpentines, pillow lavas, and radiolarian chert—always in that order, upsection. Now, in Marin, he said to Lawson, "These rocks are the same." There were solid cliffs of red chert, and pillow lavas as well. On the San Francisco side of the strait was a headland of serpentine. Steinmann commented that since the chert was stratigraphically at the top of the sequence the whole assembly must have come from deep sea. In 1905, Steinmann published a definitive study of the same three rock types threading through the Alps. They became known in the science as the Steinmann Trinity.

In 1838, the Scottish philosopher Thomas Dick, of County Angus, had published his "Celestial Scenery; or, The Wonders of the Planetary System Displayed; Illustrating the Perfections of the Deity and a Plurality of Worlds," in which the movements on the earth's surface that eventually became known as continental drift and plate tectonics were—as far as is known—first proposed. Dick noted how neatly western Africa could lock itself tight around the horn of Brazil, "and Nova Scotia and Newfoundland would block up a portion of the Bay of Biscay and the English Channel, while Great Britain and Ireland would block up the entrance to Davis's Straits." Such an assembly would "form one compact continent." And "a consideration of these circumstances renders it not altogether improbable that these continents were originally conjoined, and that at some former physical revolution or catastrophe they may have been rent asunder by some tremendous power, when the waters of the ocean rushed in between them, and left them separated as we now behold them." I am indebted to Alan Goodacre, of the Geological Survey of Canada, for this high-assay nugget, which he reported in a letter to *Nature* in 1991, backdating by

three-quarters of a century the continental-drift hypothesis attributed in textbooks to the German meteo-
rologist Alfred Wegener. Of the two, Dick fared better,
for while his proposition achieved no significant attention, Wegener's won a considerable fame that rapidly
decayed into notoriety. In 1915, Wegener published
"Die Entstehung der Kontinente und Ozeane," in
which he based his concept not only on the jigsaw fit
of Africa and the Americas but also on the likeness of
certain rocks on the two sides of the ocean, and on
comparisons of living and fossil creatures. For half a
century, Wegener—in life and in death—was a target of
scorn. His idea provoked gibes, jeers, sneers, derision,
raillery, burlesque, mockery, irony, satire, and sarcasm,
but it could not be ignored. In 1928, the American Association of Petroleum Geologists published a sympo-
sium on continental drift. It included a paper called
"Some of the Objections to Wegener's Theory," by
Rollin T. Chamberlin, of the University of Chicago,
who expressed what was then the prevailing attitude
among geologists and would continue to be until the
nineteen-seventies, after which it would cease to pre-
vail but not to survive:

Wegener's hypothesis in general is of the foot-loose type,
in that it takes considerable liberty with our globe, and is less
bound by restrictions or tied down by awkward, ugly facts
than most of its rival theories. Its appeal seems to lie in the
fact that it plays a game in which there are few restrictive
rules and no sharply drawn code of conduct. So a lot of things
go easily. But taking the situation as it now is, we must either

modify radically most of the present rules of the geological game or else pass the hypothesis by. The best characterization of the hypothesis which I have heard was a remark made at the 1922 meeting of the Geological Society of America at Ann Arbor. It was this: "If we are to believe Wegener's hypothesis we must forget everything which has been learned in the last 70 years and start all over again."

No one, for the time being, connected the Steinmann Trinity to Wegener's drift. For about four decades, the two ideas hung on to the trailing edge of the science, with no one suspecting that Wegener's idea would effloresce as a scientific paradigm, or that the ophiolite story (the advanced edition of the Steinmann Trinity) would, thereafter, provide the chapter headings to the plate-tectonic history of the world.

There were hints. In 1936, Harry Hess, of Princeton, whose "History of Ocean Basins" would—in 1960—introduce the spreading seafloor and begin the new tectonic story, gave a paper in Moscow in which he related Alpine peridotites to island arcs and called this "a contribution to the ophiolite problem." The ophiolite problem was manifold. Most geologists did not accept the association of such disparate rocks. The few who did asked elemental and unanswerable questions: Are the rocks in their original place? If not, where have they come from and how did they move? Peridotite, which in alteration becomes serpentine, is now seen to be of the deepest origin of any rock found at the earth's surface. It is now thought to be rock of the earth's mantle, and few disagree. Hess had studied two belts of Alpine-

type peridotites in the Appalachians. He asserted that
they were not just garden-variety igneous rocks that
had come up as magma under existing Appalachians
but rock that had intruded much earlier, on the edges
of a geosyncline. Hess went on to say that peridotites
seemed to be introduced in the initial phases of mountain building—and not thereafter. He said they were
cold, intact, and solid as they came up in the rising
mountains—that they had been, in other words, tectonically emplaced. Hess was describing a collision between a continent and an oceanic subduction zone, but
in 1936 he didn't know it. In Moores' words, "This is
the clearest example I know of of a guy saying the right
thing for the wrong reason. Hess said this was the most
important event in the forming of the Appalachians,
this 'intrusion' in the Ordovician. It was. But plate tectonics did it."

By 1955, the burst of data that had come with the
ocean-exploration programs of the Cold War had
yielded—in the work of Russell Raitt, of the Scripps In-
stitution of Oceanography, and Maurice Ewing, of Columbia University—the first descriptions of ocean crust
to be derived from seismic refraction. Everywhere, the
ocean crust seemed to be some sort of package—a once
molten but nonetheless zoned assembly, in three general bands.

The German geologist W. P. de Roever was the
first to imagine rock of the earth's mantle in the thin
air of the Alps. In a 1957 paper, he said that the Alpine
peridotites appear to be solid intrusions; they have deformed fabrics; they appear to be coming up from the

mantle. For "solid intrusions" you could read "emplacements from elsewhere." In their deformed fabrics you
could see that they had been moved. How they had
been moved remained unclear.

In 1959, Jan Brunn, who had mapped the Vourinos
ophiolitic complex in Macedonia, published an abstract
in French in which he compared ophiolites to the Mid-
Atlantic Ridge. He was the first person ever to suggest
that the ophiolites of the aerial world were similar to
crust found at mid-ocean ridges. He compared ophiolitic rocks to dredged rocks. All this was before seafloor spreading was recognized, and no one noticed the
work of Brunn. But it was Brunn who took the Steinmann Trinity and the ophiolitic sequence out of the Oz
of geosynclines and placed them in the center of the
widening oceans.

Over the next nine years (1960–68) appeared the
twenty-odd scientific papers reporting the plate-
tectonics story: that plates are essentially rigid, and deform at their boundaries; that all plates include ocean
crust, and generally a very large amount of it (the continents are passengers on the plates); that new seafloor
moves away from a spreading center until it goes down
into a trench to be consumed; that plates sliding past
each other (as at San Francisco) do so in strike-slip
sporadic jumps; that ocean crust colliding with continental crust can pry up something like the Andes; that
continental crust hitting continental crust will build
Himalayas, Urals, Appalachians, and Alps.

While these novel facts were still for the most part
unknown, Moores, in Macedonia in the early nineteen-

sixties, was becoming thoroughly at home with the petrology and structure of the Vourinos Complex, about thirty miles west of Mt. Olympus. Brunn's ideas notwithstanding, Moores thought of the Vourinos rocks as homegrown—in his words, "a partially molten diapiric blob." Diapirs are bodies of rock shaped more or less like hot-air balloons, and balloonlike they rise, crashing their way into the country rock above them. Harry Hess, one of Moores' supervisors, went to Greece to inspect his work. Hess was already in the process of abandoning the geosyncline and shedding the Old Geology like old skin. He decided that the Vourinos rock had formed in an oceanic environment, and probably at a spreading center. Moores, clinging to what he had been taught (by, among others, Hess), decided that Hess was crazy.

Moores' conservatism can be understood in the light of the disregard in which ophiolites were generally held in the United States at that time. When a graduate student at Stanford, doing field work not far from the university, suggested that the local sediments had been deposited on an ophiolitic complex, the Geology Department specifically forbade him to use the word "ophiolite" in a Ph.D. thesis. The professors explained that ophiolites were a wild European idea, clearly wrong—and, in any case, not applicable to California.

The concept of the spreading seafloor had gained acceptance by 1966, when Moores first saw the geologic map of Cyprus. Moores and Vine prepared to go there, but had to wait, because of the political tension in the six-year-old country. They worked there in 1968 and 1969. The evidence was convincing that Cyprus was essentially a piece of ocean crust, thrust up in some way and now aerially exposed. Moores and Vine's paper, which has influenced all subsequent understanding of ocean crust, was the first to establish ophiolites as ocean-floor remnants, for the most part formed by spreading processes. This odd collection of rocks ranging from water-cooled lavas to mantle blocks, so difficult to explain in continental settings, could now be seen not as ordinary igneous formations but as tectonic features that had moved from one place to another in the course of epic alterations of landscape.

In Pacific Grove, California, at the end of 1969, a Penrose conference on "The Meaning of the New Global Tectonics" drew structural geologists from all over the world. William Dickinson, of Stanford, dismantled the geosyncline and assigned its parts to various aspects of plate tectonics—collisions, island arcs, abyssal plains, mélanges, trenches, transform faults. Moores describes the conference as "a watershed of geology—that was when people really began to realize how important plate tectonics was." Listening to Dickinson, he thought of all the ophiolitic and volcanic-island rock that he was seeing in the Sierra Nevada and the Coast Ranges, and it occurred to him that these mountain systems could be understood in terms of island arcs accreting. The arrival times of ophiolites could date successive mountain-building events. In his words, "Recognition of the fact that ophiolite emplacement had to be by the collision process meant that you could explain the western part of the United States by that kind of sequence. That idea came to me on the last morning of the Penrose conference in 1969, and I wrote it down: the idea that you could explain the progressive eugeosyncline-miogeosyncline development and the progressive orogenies that you see in western North America as a series of island-arc complexes—the things we call terranes—that have collided with the continent. I was so excited I could hardly sit still for several days—in fact, for a couple of weeks after that. I came back to Davis all bubbling over." Before long, he had sent a paper to *Nature*. It appeared in 1970 and was the first to suggest the collisional assembling of California and of vast related portions of the North American Plate.

The idea that California is in large part a collection and compaction of oceanic islands was a reverberation of an ancient myth as well as a development in a science. For at least two thousand years, people described certain undiscovered islands with a force of imagination that became belief. In the Middle Ages and the Renaissance, such islands appeared on global maps, and when navigation revealed that they were not where they were said to be cartographers moved the islands to new locations in unexplored seas—the Fortunate Isles, for example, and the Seven Cities of Cíbola, and the Lost Atlantis. One such island was California, a utopia of the western Atlantic Ocean. It is described as follows in *Las Sergas de Esplandián*, a Spanish romance published in 1508:

Know, then, that, on the right hand of the Indies, there is an island called California, very close to the side of the Terrestrial Paradise, and it was peopled by black women, without any man among them, for they lived in the fashion of Amazons. They were of strong and hardy bodies, of ardent courage and great force. Their island was the strongest in all the world, with its steep cliffs and rocky shores. Their arms were all of gold, and so was the harness of the wild beasts which they tamed and rode. For, in the whole island, there was no metal but gold.

The wild beasts were griffins—half lion, half eagle—which the women rode through the air into battle, and which they trapped as fledglings. To the griffins they fed the voyaging men who came their way, and their own male infants. Ruling California was the "mighty Queen, Calafia . . . the most beautiful of all of them, of blooming years." (Translation by Edward Everett Hale, 1864.)

With an eye on the new tectonics, the paleontologist James W. Valentine, of the University of California, Davis, worked out a curve of the distribution of marine-invertebrate families across Phanerozoic time (the past five hundred and forty-four million years). He saw the diversity of creatures expand, decline, rise again. The thought occurred to him that if numerous small continents were spread around the world in fair proximity to the equator one would expect the diversity of life to be very high, whereas if continental masses should happen to be clustered (and especially if they were clustered around a pole) diversity should be very low. A typical Valentine graph showed creatures from continental shelves starting out at a very low level of diversity in late

Precambrian and early Cambrian time, coming up to a high level in mid-Paleozoic time, then crashing at the end of the Permian, and then rising again. Valentine showed his graphs to Moores, and said, "I wonder if you can explain these patterns of diversity in terms of continental drift and redistribution of land."

Among the results of this dialogue were papers by Valentine and Moores in *Nature* (1970) and the *Journal of Geology* (1972). The title of the second one was "Global Tectonics and the Fossil Record." The concept of continental drift had always implied the preexistence of a supercontinent, which Wegener had called Pangaea. After all, if Australia and Africa and the Americas and Eurasia spread apart from their obvious fit, they had to have been together in the first place. The first place—according to the newly determined vectors of the lithospheric plates—was two hundred million years ago, when Pangaea began to split into Laurasia and Gondwanaland. In the second place, they split farther, to sketch the present globe. Valentine's diversity patterns were in harmony with this story: the greater the breakup of landmasses, the more diverse the fossil families. Moores looked at mountain belts that had come into existence hundreds of millions of years before the dispersal of Pangaea. If the new theory worked, it would work not only forward in time but backward. Gradually, he reassembled Paleozoic and Precambrian continents—not only the continents that came together to create Pangaea but also the continents that came together to create an earlier supercontinent, which no one had thought of before. Moores and Valentine called it Protopangaea, or Pangaea Minus One. More widely, it has become known in the science as Rodinia (from a Russian word meaning "motherland"). Since agglomerations and dispersals of terrane seemed to be cyclical in nature, there may have been who knows how many supercontinents before Rodinia. The new theory was like a stable and inventive structure built in the mind of a composer in advance of a composition, but there were those who didn't like what they heard. "Global Tectonics and the Fossil Record" was attempting to demonstrate how continental drift affects evolution, and—heaven knows—it succeeded in enraging no small number of geologists and paleontologists, who felt that Moores, especially, and other "plate-tectonics boys" were ignoring phenomena like ocean currents in their headlong lust to write all aspects of the geologic pageant into the plate-tectonic model. But in 1978 Patrick Morel and Ted Irving, of Canada's Department of Energy, Mines, and Resources, presented paleomagnetic evidence for Pangaea Minus One.

Moores and Valentine also sensed a relationship between plate-tectonic history and the history of the level of the sea. Moores explained, "If you look at the stratigraphic record on platforms such as the midcontinent of the United States, you see times of high stands of the seas, when the continent was nearly submerged, and times of really low stands of the seas. Could that be related to continental drift? Take an average ocean basin and add a hot and voluminous spreading ridge. You will diminish the volume of the ocean basin and force water up over the continents. Conversely, if ridges

die for some reason—lose their heat and collapse, or
otherwise disappear—that will increase the volume of
ocean basins and allow seawater to drain off the continents. The ocean's transgressions and regressions seem
to represent seafloor spreading going on or not going
on. Others worked on this before we did, but we extended it back to the Cambrian-Precambrian boundary.
In the geologic record, you see a great regression of
seas in late Precambrian time, then transgression in
Cambrian and Ordovician time, then regression in
Permo-Triassic time, and then the transgression of
Cretaceous time. The late-Precambrian regression coin-
cides with Pangaea Minus One. As it split apart, with
spreading centers forming all over the place, the
Cambro-Ordovician transgression occurred, and when
the smaller continents came back together there was
the Permo-Triassic regression, and when they split
apart again came the famous Cretaceous transgression,
when Colorado was underwater."

Hugh Davies, of the Papuan Geologic Survey,
published cross sections of New Guinea that showed
a huge ophiolite dipping northward. The Australian
Plate, like a shovel, had lifted this piece of Pacific mantle and crust. The ophiolitologists Robert Stevens, John
Malpas, and Harold Williams, of Memorial University
of Newfoundland, described a series of ophiolites in
Newfoundland as remnants of an arc that collided
with North America in Ordovician time, signing in the
Taconic Orogeny. Eli Silver, of the University of California, Santa Cruz, working in Indonesia, traced a large
ophiolite there from its emplaced position on a sub-

merged microcontinent northward into an ocean basin.

As comparable research went on around the world, a question that attended all of it was "What ancient geography can be deduced from ophiolites?" and spectacular deductions continued to be forthcoming. The ophiolite suite seemed not only to spell out in detail the process of formation of oceanic lithosphere but to record plate collisions of which all other evidence was long since gone. Vanished oceans were recalled, and vanished plates inferred, as continents were deconstructed and continents were reconstructed. An ocean (or oceans) that once existed where the Atlantic is now had closed from north to south in Cambro-Devonian time, and the Permian disappearance of the ocean where the Urals are now had completed Pangaea. The Pacific Plate, now the largest in the world, did not exist then, but a spreading ridge, propagating westward, split Pangaea into Laurasia and Gondwanaland in late Triassic and early Jurassic time, and opened the Tethys Ocean. The early central Atlantic was a part of Tethys. As oceans came and went and continents evolved, island arc after island arc had been swept into larger masses— a story that could suggest that the first dry lands of Genesis were arcs accreting in a globe-girdling sea.

As with so many things that were obscure or mystifying before the arrival of plate tectonics, the discovery of the origin of ophiolites was something like the discovery that the rock you have been using for twenty-five years as a doorstop is actually the Stone of Scone. Suddenly, for example, the concept unmentionable in a Stanford doctoral thesis was helping to tell the story of California as it had not been understood before. Moores, reflecting on this, said to me once, "If the story of California sounds fantastic, with all its accreting arcs and mélanges coming from the western sea, just look at a map of the southwest Pacific—look at the relationship between Australia and Indonesia right now."

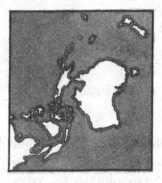

As fresh attention accrued to the ophiolitic suite where—in completeness—it was found on land, the actual dimensions of the ocean lithosphere could be measured part by part from top to bottom, clarifying the results of seismic refraction. After the sequence has formed at spreading centers—where it is in large part liquid and is much swollen with heat—it cools and thins as it moves away, and after travelling twenty million years and seven hundred miles is a deep cold slab:

A few tens of metres of ocean sediments drift down upon the deep cold slab, settling on top of

a kilometre or so of pillow lavas, under which is

a kilometre or so of sheeted dikes, under which is

a kilometre or so of plutonic rock (plagiogranite, gabbro), under which is

a kilometre or so of plutonic rock in which cumulate crystals settled in layers upon a distinct chamber bottom—

the Moho—

under which is a kilometre or so of mantle rock, some of which was melted in the spreading center and some of which is peridotite in its solid original form and if water has reached it is serpentine.

When Moores talks about ophiolites with fifth-graders in Davis, he sketches the ocean floor much as he has for me, simplifying the vertical sequence and presenting an idealized rock column that closely resembles the one above. It is more than just a useful model. In its general way, it is accurate. But as a description it is only somewhat more encompassing than to say that Herman Melville wrote a novel about a one-legged madman in vengeful pursuit of a whale. Remarkable as ocean lithosphere may be for its worldwide youth and repeated character, it is not nearly as simple as that summary outline. In Moores' words, "Where ophiolites are made, in spreading centers, hot fluids are mixing, cooling, and so forth. We are not talking about clear sedimentary layers; ophiolite contacts are gradational. Moreover, there are various types of ophiolites: some are from the basins behind or in front of island arcs, some from the intersections of spreading centers and transform faults. There is undersea 'weathering.' Parts of the sequence erode, more sediment comes down, and as a result there are hiatuses in seafloor rock, just as there are in rock that accumulates on land. In Italian ophiolites, the diabase is missing, and so is the gabbro. The serpentine is full of calcite—it's called ophicalcite, very beautiful white and green and red stone, expensive building stone. There is some gabbro in Elba. But there are no sheeted-dike complexes in Italy. Obviously, Italian ophiolites formed in a different ocean environment, and what that may have been is not well understood. Oceanic crust is not a simple three-layered thing, as geophysics is telling us now. Geophysicists are unable to produce a consistent model. Nature is messy."

One day, on a field trip we made to Cyprus, Moores did a long and detailed inspection of an outcrop he had not studied before, and figured out a chain of magmatic events in which layered gabbro had come first and a plagiogranite sill had intruded below the gabbro—an inversion of the usual sequence. "This reminds us not to take the chart of the ophiolites simply," he said. "Layered gabbro may be lower than plagiogranite in the master chart, but here we see a plagiogranite sill under the layered gabbro. It came in later. Things don't always happen in the earth as they do on charts."

Not to confuse me but just to give me a reality shot, he sketched out what he described as "an expanded ophiolitic assemblage"—an elaboration of the picture I had written on my palm. He repeated the generalized column with enough added detail to suggest the actual complexity of the rock that lies below the oceans. I could forget it as soon as I read it, but at least I ought to sense the tangle of nature, and thus the nature of science.

Where you find ophiolites on land, you might find at the top of the sequence the shallow-water limestones of the sea from which they emerged, or even laterites from soils formed in air as the ophiolite was lifted by the continental margin.

The deep-water sediments that drifted down upon the moving lithospheric plate may be chalk (as in Cyprus) or the product of volcanoes (as in the Smartville Block) or chert (as in Italy or Greece). They tell you something about the oceanic environment through which the lithosphere travelled.

Beneath the massive pillow lavas are

more pillow lavas, shot through with diabase dikes that came up in molten state with enough pressure to continue past the

zone of massive sheeted dikes. If the ophiolite were an animal, these originally vertical laminations would be its brain. Taken together, they are the tape measure and chronometer of the seafloor. As each new dike forced its way into the complex, the seafloor spread that much. New dikes would intrude every fifty to a hundred years, often splitting the previous dike up the middle. The average width of the split dikes would be about seventy centimetres. They recorded absolutely the episodic widening of the ocean lithosphere, in contrast to the arithmetical notion of geophysicists, who suggest that seafloors spread continuously at a quotient number of centimetres per year.

Among the plagiogranites are large diabase dikes (feeders of the sheeted complex)

and large diabase dikes are among the massive varitextured gabbros as well.

The stratiform gabbros are generally cyclic: plagioclase and pyroxene cumulate above olivine and pyroxene cumulate with intercumulate plagioclase (and sometimes olivine and plagioclase) above olivine cumulate with some chromite in it, a combination known as dunite. You will not at once recognize all these things where an ophiolite has made an appearance in a roadcut or a mountain cliff, but at least try to remember that somewhere within this zone is the geophysical Moho—

and at the bottom of this zone is the petrologic Moho.

I have to interrupt him. Two Mohos? How could there be two Mohos? *The* Moho, as fifth-graders can tell you, is where the crust ends and the mantle begins—about five kilometres down from the ocean bottom and thirty-five kilometres below most places on continents and as much as sixty kilometres below deeply floating mountains. "Moho" is a geophysical term, coined in honor of the Croatian seismologist Andrija Mohorivičić, who in 1909 discovered the crust-mantle boundary. When geophysicists examine their autodriven strip-chart recordings, which look like close-up photographs of matted gray hair, they see what Mohorivičić saw. They see seismic waves speeding up when they hit the olivine-rich cumulates, and they call that the change from crust to mantle. Geologists looking at ophiolites in geographical settings see mantle material below the olivine-rich cumulates and discern that the mantle supplied the olivine that went into the rock above it and now devilishly accelerates the geophysicists' seismic waves. Below the olivine-rich cumulates geologists see what they regard as the true transition from crust to mantle, and they call it the petrologic Moho. Geophysicists insist their machines can't be wrong, the nature of the rock notwithstanding. Moores says to remember, however, that Moho means "Mohorivičić discontinuity" and the seismic discontinuity is decidedly where geophysicists say it is. The discontinuity is seismic and is recorded on paper, but the crust-mantle boundary, which is lower, is recorded in the rock. Nature, in this case, is not messy or confused. The science is—for the time being. Like the early cartographers piecing together the face of the globe, geologists and geophysicists are now trying to map places that no human being will ever see but which are features of the earth no less than Scotts Bluff or the Shetland Islands, and were features of the earth when Scotts Bluff and the Shetland Islands were other rock in other places, and will be features of the earth when Scotts Bluff has totally disintegrated and the Shetland Islands are under the sea. The two Mohos are like a camera's divided range finder trying to close. The two Mohos are an imperfectly mapped frontier

under which, in the ophiolitic sequence, the kilometre or so of mantle rock is peridotite, a general term for rock composed mainly of olivine and a little pyroxene. Depending on the kind of pyroxene and the amount of pyroxene that is in it, peridotite is also called harzburgite and lherzolite and dunite, and, if water has reached it, serpentine. Peridotite that has moved in its solid original form is called tectonite, and presumably extends to the bottom of the lithospheric plate. One of the thickest measurable sections of mantle rock emplaced on any continent today is in Macedonia, and is about seven kilometres from top to bottom, a considerable weight to lift from the mantle into the air.

In this young field within earth science, the two Mohos are scarcely the sole battleground. Because ophiolites develop not only at mid-ocean ridges but also in the small spreading centers associated with island arcs and sometimes along oceanic transform faults, arguments about their origins can be intense. There is considerable agreement that the Smartville ophiolites relate to the arrival of an exotic island arc, and that the Bay of Islands ophiolite complex, in Newfoundland, formed at a mid-ocean spreading center. But Moores thinks Cyprus formed in mid-ocean, and most ophiolitologists do not. Italy's ophiolites, with their missing parts, seem to some, but not others, to be fragments of oceanic transform faults. The Papuan ophiolites of New Guinea are so complicated that they seem—to some workers—to have come not only from a mid-ocean spreading center but also from behind an island arc.

Moores says that ophiolites are more important as models for the mechanism of spreading than they are as relics of the environments where they were made. The elapsed time between formation and emplacement is measurable by various methods, and some people argue that emplacement seems to have happened too quickly—an average of thirty million years—for most of the world's ophiolites to have derived from mid-ocean centers. Moores says he has no argument with people who hold that most ophiolites form near continents—in fore-arcs or back-arcs—but he insists that there is enough time to bring them in from mid-ocean ridges as well. Geologists argue about the chambers of magma under spreading centers, and whether the chambers were continuous over time or were punctuated units that crystallized and were followed by new chambers that in turn crystallized, and so forth. They wonder, above all, why only a few ophiolites are more than a thousand million years old, while the earth itself is four and a half times older. For the last thousand million years they can work out the tectonic history—the shifting shapes of continents, the rise of long-gone mountains—from the ophiolites that were left behind. Before that, in the early Proterozoic and the Archean and Hadean eons, what was going on? Was something different going on? Something other than plate tectonics?

The long argument over how ophiolites are emplaced has provided workers with a more immediate distraction. In 1971, R. G. Coleman, of the United States Geological Survey, proposed that where ocean crust slides into a trench and goes under a continent, a part of the crust—i.e., an ophiolite—is shaved off the top and ends up on the lip of the continent. He called this "obduction." In 1976, David Elliott, of Johns Hopkins, decided that rocks could not stand the roughshod tectonics that Coleman had proposed—that the ophiolites would be shattered into countless parts and would just not make it up there. Elliott proposed gravity sliding—ophiolites as toboggans coming to rest. But something would have to lift the seafloor and break some off before it could slide. And what, for example, could have lifted the Macedonian mantle more than forty-five thousand vertical feet? The idea that enjoys the widest acceptance was around before either of the others, but no one much noticed, because it came from a graduate student and a postdoctoral fellow. In 1969, Peter Temple and Jay Zimmerman had proposed that the emplacement of an ophiolite might occur when a continental margin goes down under ocean crust, jams a trench, and then isostatically lifts the ocean crust.

Their proposal, derived from seismic data, recognized that the subduction of lithospheric plates was far more varied than people had supposed. Not only did ocean floor dive under continents but also—and much more commonly—it dived under other ocean floor, like two carpets overlapping. The lower slab, after melting, rose through the upper one as a volcanic-island arc. Now the island arc begins to move with the plate on which it rests. Plate motions shift. New trenches form. In back-arc basins, new ocean crust is made. Some island arcs go in one direction for a while and then reverse themselves. They choke a trench, say, and then go the other way, eating up their own crust as they go. Some of the crust might get emplaced as an ophiolite. The Marianas back-arc basin is spreading now, and so is the Lau-Havre Basin behind the Tonga-Kermadec arc, and so is the basin behind the South Sandwich Islands. Between Indonesia and the Philippines are two trenches that are eating their way toward each other and if nothing stops them will destroy each other. Some people—Moores among them—think that in similar fashion off Jurassic California two trenches were active simultaneously, the easterly one dipping to the east and the westerly one dipping west, and both were destroyed during the Smartville emplacement. In *Geology* for December, 1983, the volcanologist Alex McBirney, after dealing with the increasingly complicated attempts to relate igneous rocks to plate tectonics, closed with a vision of the decade to come. He said, "I predict that our present confusion about igneous rocks will rise to undreamed-of levels of sophistication."

Within half a decade, it was decided that the Smartville arc had formed where a spreading center developed in a transform fault or a fracture zone, after which the vector of the plate changed. When plate motions change, transform faults may turn into subduction zones or spreading centers. Forty-three million years ago, for example, the Pacific Plate changed its heading from north to northwest. In the hot-spot track of the Emperor and Hawaiian seamounts, the change is recorded in a pronounced bend—from north-south to northwest-southeast—dating exactly to that time. As transform faults turned into trenches all over the ancestral Pacific, the Tonga-Kermadec arc was created, and the Aleutian Islands and the Marianas. Something like this seems to have happened a hundred and sixty million years ago, creating the Smartville island arc.

That, Moores said, is where things are today—in

the understanding of an ophiolite and its history in the world. If I was confused, so were geologists, not to mention geophysicists. "It has taken people a lot of time to stop thinking of these things as locally derived igneous rocks, and to begin thinking of them as transported tectonic features." If the Smartville Block was bewildering in the third and fourth dimension, I should at least reflect on the complexity of the surface, where Yuba City is the county seat of Sutter County, Marysville is the county seat of Yuba County, Auburn is the county seat of Placer County, Placerville is the county seat of El Dorado County, and El Dorado is the county seat of nowhere.

Heaping Pelion upon Ossa, Moores reminded me that there was a possible ophiolite higher in the Sierra and that in many ways it perplexed the relative simplicity of the story he had been telling. Known in geology as the Feather River peridotite, it was a large body of serpentine and related rocks that we had seen near the interstate just below Dutch Flat, where we went into a canyon one day whose walls, soft to the touch, were snakeskin black and green. The Feather River peridotite defied explanation, because it was much older than the terranes on either side of it. If this region of the western United States consisted of accreted terranes, progressively younger as you moved west, what was the Feather River peridotite doing there between Sonomia and Smartville? "The dates on it range from the Devonian through the Permian," Moores said. "We don't know how it got here. It was metamorphosed at six hundred degrees centigrade and twenty kilometres down. Must have been an extraordinary thrust that brought it up from twenty kilometres. Some of the rocks it deforms on its eastern contact are Triassic in age. It is older than Sonomia but was emplaced later. It is older than Smartville and lies east of Smartville, but—who knows how or why?—it was emplaced later. We have the story of several successive terranes neatly tying onto North America, and then we discover that the Feather River peridotite is between No. 2 and No. 3, that it is much older than any rock in No. 2, and that its emplacement is younger than the arrival of No. 3. The Feather River peridotite was about two hundred million years old when emplaced here. What does that mean? That it's anomalously old? That it was picked up first by an island arc and ultimately emplaced twice? Is it, in the first place, an ophiolite? What else can it be? It includes serpentine, unserpentinized peridotite, metagabbros, other amphibolites of diverse parentage, and, possibly, deformed sheeted dikes. What does that suggest?"

To Steve Edelman, a young structural geologist and tectonicist trained at Davis and the University of South Carolina, that suggested a genuine ophiolite, however baffling its history might be, and he believed he had found the sheeted dikes that would seal the discussion. He was the one geologist who had ever come out of the Sierra mentioning this possibility. One day in 1989, Edelman got off a train in Davis (being between jobs and between grants, he could not afford to fly) and started for the mountains to enrich his field research. He asked Moores to go with him, and I went along.

Edelman, whose beard is red, looked like a tennis

player prepared to serve with a rock hammer. He was wearing a pink eyeshade in foam plastic, an aqua T-shirt, shorts, Adidas shoes, and white socks with red and blue bands. We went down a very steep canyonside more than five hundred vertical feet into the narrow defile of Slate Creek, close to the mining camps of Grass Flat, French Camp, and Yankee Hill, and closer to Devil's Gate. Slate Creek was a place of spare light, where miners had removed a hundred thousand ounces of gold. In California canyons as remote as this, gold has more recently been grown in the form of plants with spiky leaves of the sort that were painted by Henri Rousseau. In a statewide effort to stop the industry, the government in Sacramento had organized something called CAMP—Campaign Against Marijuana Production. CAMP narcs were said to be posing as geologists when they roamed the wild country. A geology graduate student from the University of Texas, doing field work in the northern Coast Ranges, had been murdered, presumably by marijuana growers—shot in the back of the head. The killers have never been caught. A man in the California Division of Mines and Geology whose work frequently takes him into the Sierra foothills goes into every bar in every old mining camp and forest hamlet within a radius of five or ten miles and tells everyone present what he is doing and where. Out in the crops and outcrops, when he encounters people they know about him.

To the nineteenth-century miners a lot of rock was slate. Slate Creek was flowing over a beautiful gray diabase. I remarked that it was one of the clearest streams

I had seen anywhere south of the Brooks Range. Moores marvelled, too, saying that he could see "shear-sense indication in porphyroclasts at the bottom of the creek—it's that clear."

This seemed to please Edelman, but not greatly. He had asked Moores to come to Slate Creek to see if he could agree that the diabase, as old and recrystallized and murkily deformed as it was, showed any pentimento of the laminations and chilled margins of sheeted ophiolitic dikes.

Seeing nothing in his lens that impressed him as he moved from ledge to ledge upstream, Moores struggled to be cooperative. He said, "It's not a good stretch. We have to get around the next bend. . . . I think I see some folding, some layering. Maybe it's gabbroic."

Edelman seemed to speed up a bit.

"Maybe that is a pillow there," Moores said, hopefully. But doubt was the pillowcase.

Now Edelman was on the run. He promised better things ahead. We had not yet reached the exposures he wanted us to see. Around the next bend, he closely scrutinized a big outcrop that jutted into the stream. He said that in his judgment it was a set of sheeted dikes. What did Moores think?

Moores leaned in to the rock, lingered, drew back, and said, "It's O.K. if you're a believer. It's not a skeptic's outcrop."

"And this one?"

"Only slightly more convincing than the last."

Edelman was right, though. The farther we moved

upstream, the more the gray rock seemed to exhibit the features he had asked Moores to witness. But the weathering and deformation were such that the laminations were not easy to discern. Moores, with his leather cases and leather pouches lined up on his belt, his wide-brimmed fedora shading his beard, appeared to have been there since 1849.

"I don't think a Howard Day type skeptic is going to believe this," he said, mentioning a metamorphic petrologist at Davis.

"No," Edelman said. "I guess not."

Moores said, "Fortunately, there's only one of those in the world."

Edelman said, "I know it gets good up here ahead of us. Suspend your judgment until you see the good stuff—until you see the well-defined sheeted dikes."

A little farther on, Moores paused at an outcrop for a long close look, with and without a lens. Minutes went by. Slate Creek went by, making the only sound. At last, Moores said, "That's a good one. That would convince Howard Day."

The next ledge was even better. Moores examined it for some time. Edelman's expression was full of sniffed victory. Moores said, "I have no doubt that that is a chilled margin. And that. And that." Looking into Edelman's eyes, he added, "But I'm a believer."

Another few yards, another rock face, and Moores said, "If you were an ophiolite and someone took you down to six hundred degrees and twenty kilometres, maybe this is what you would look like. If this is not a dike complex, what else could it have been? Nothing, is my answer."

Edelman said, "So. Do you believe in the Devil's Gate Ophiolite?"

Moores said, "I believe in the Devil's Gate Sheeted Dike Complex."

Edelman: "You happy?"

Moores: "Yes."

Edelman: "Another ophiolite in the Sierra."

A lot of geology is learned now from seismic waves and satellites, and pieced together on printout paper in artificial light. Neither a seismometer nor a satellite was ever going to see what Edelman had seen.

Moores said, "If you took Japan and its old ocean crust and collided it with the state of Washington, and the old crust was shoved over Washington, you might have something like the Feather River peridotite. If this is an ophiolite, it is bigger than the Troodos of Cyprus. It does not appear to be a part of Sonomia. It could be a subsidiary of Smartville. You just don't know what it's doing here. You just know it's big and it's important."

If the Feather River peridotite inconveniences the exotic-terrane story, it is well that it should, he said. The old picture of the western margin of North America is gone, but not long gone. The present description of assembling terranes is so new—and calls for so much working out—that it hardly requires acid-free paper. Did a large pre-assembled terrane—the Stikine Super-terrane, of which Smartville would be a part—dock in the Jurassic? Or did the United Plates of America, as they have been called, arrive separately? "Most of the ophiolites from the Brooks Range on down through central British Columbia into the western United States and in Baja California and Costa Rica are Mesozoic in age," Moores said. "They appear to represent some sort of island-arc complex that collided with western North America in mid-Jurassic time. Singly or collectively? That's hard to say. You can make it the one or the other. We can't work that out now. The timing seems to be different. It seems to be lower Cretaceous in the Brooks Range, mid-Jurassic in British Columbia, mid-Jurassic in the Sierra Nevada, and somewhat younger as one goes farther south. That could be a single, ragged-edged collision. Or it could be several terranes coming in. We're not prepared to say."

There remain in the world, of course, geologists who are not prepared to say that exotic terranes exist in such prodigious quantity. Homicidal in their sarcasm, they still like to assert that their less conservative colleagues are prone to name new terranes for any change of lithology, at any formation boundary—in fact, at any place, however small, where it is easier to claim that something is exotic than to figure out the relationships of present and missing parts. In the geology of such people, it is said, a microterrane is a field area. A nano-terrane is an outcrop. A picoterrane is a hand specimen. A femtoterrane is a thin section.

Although the outer shell of two-thirds of the earth is rock of the ocean crust, it is so inaccessible to the field geologist that to study even the fragments that have broken off on continental margins requires prodigious travel. Moores has pursued his specialty from Oman to Yap to Tierra del Fuego to Pakistan, and routinely has returned to the eastern Mediterranean—most of all to Cyprus. One autumn in the nineteen-eighties, when I was working in Switzerland, I flew to Cyprus to watch him do his geology there.

He met me at Larnaca, we drove north, and within the hour were tectonically deconstructing a huge broiled fish and drinking dark Cypriot wine. Politically, Cyprus was in Asia, Moores said, but geologically it belonged to no continent. It rested on the lip of Africa but was not African. It was not Eurasian. In the lowercase and literal sense, it was mediterranean. Long after the last supercontinent began rifting and its new internal shorelines bordered the Tethys Ocean, what was to become the foundation rock of Cyprus welled up as magma in a Tethyan spreading center. At that time, about ninety million years before the present, the Eurasian and the African sides of Tethys were twice as far apart as they are now. They continued to separate for another ten million years, and then plate motions changed. As the North Atlantic began to open, Africa began to move northeast, closing with Eurasia, as it continues to do. Ongoing results include the Alps, the Carpathians, the Caucasus, the Zagros—in Moores' words, "a big slug of deformation that's throwing up mountains everywhere." By late Miocene time, geologically near the present, not much remained of Tethys in that part of the world except the Mediterranean Sea, the Black Sea, and the southern Caspian Sea.

It was in the late Miocene that Africa levered the Tethyan floor, and broke off Cyprus. The ophiolite was thickly covered with chalk, which had settled upon the pillow lavas as a clean lime ooze, unadulterated with sediment from any landmass—as clear an indication as you could ever hope to find that Cyprus was a piece of remote deep ocean.

For seeing and touching ocean crust—for leaning against it with a hand lens, for removing small cores to study their remanent magnetism, for mapping the varying rocks from zone to zone—there was no example in the world as well preserved as Cyprus. Shaped like a razor clam, the island had a long foot that reached up in the general direction of Turkey, which was fifty miles away. This northeastern extremity was a long low range of mountains, whose geologic history was not well understood: it seemed to be in some sense accretionary, perhaps a collection of island fragments from a confused Eurasian sea, and thus to belong—in a tectonic sense—to Turkey, which seized it by force in 1974, and established it as the Turkish Federated State of Cyprus. The Turkish Federated State of Cyprus had so far been recognized only by geologists. The thick body of the island, which rose higher than the White Mountains of New Hampshire, was the ophiolite—the exposed and integral lithospheric crust that was without continental affiliation and was the heart and substance of the independent Republic of Cyprus.

Each day, we drove out of Nicosia down the great treeless plain of the Mesaoria, on our way to the high Troodos. Trending east-west, the plain divided the island's most prominent lithologies. Off to our right about ten miles was the low silhouette of the north-coastal range, and ahead to the left were the Troodos. Through the Mesaoria ran a boundary drawn by the United Nations which was known among Greek Cypriots as the

Green Line. Turks called it the Attila Line. Along it, over the plain, ran United Nations sentry boxes, like populated fence posts, farther than the eye could see. All north-south roads, even unpaved ruts, were barricaded and marked with warnings and instructive signs: "HALT!"

The feel of day in the Mesaoria was like crouching too close to a campfire. Up in the Troodos were groves of shade. Under Aleppo pines, the air was as cool as deep water, and about as still. There were whole ridgelines of sheeted diabase, weathered out in silver blades—like thousands of playing cards in one standing deck—recording in subcenturies the spreading of ten million years. There were cliffs of chalk, and bulging extrusions of pillow basalt. There were layered cumulates in massive black gabbro. There were serpentine peaks. The highest peak was Mt. Olympus. In the Hellenic world are enough Mt. Olympuses to suggest tract housing for redundant gods. It is a godly talent that geologists have: not only to see ocean lithosphere in mountain crests but to feel comfortable in the knowledge that some of the lowest rock in the ophiolitic sequence is the highest rock of a place like Cyprus—with nothing overturned. A sawyer would also understand this—and almost anyone who could look at woodwork and see the original trees. The Cyprus ophiolite—great slab of the ocean—was bent upon the slope of Africa. It was draped, hung, arched, folded—not quite like Dali's watches, but the image would do. Water entering peridotite to make serpentine had swelled the whole affair, and then erosion had taken over, finding the serpentine within the crown of the arch and variously stripping the other stuff off the top and down the sides until the serpentine stood highest and the ophiolitic sequence (reading upward) went down the mountains in successive steps, ending in peripheral cliffs of chalk. In some of the higher country, Moores chipped away at interbanded cumulates (so-called magmatic sediments) from the deepest pools of gabbro, and among them found a trace current, which he described as "a stream channel in the magma chamber." Rapping it with his hammer, he said, "Shows you which way is up."

Among stone cabins whose metal roofs were weighted with stone, we hammered stone. We moved among ripening apples, Lombardy poplars, and redroofed white villages spread on the dry mountains. We went into deep canyons. A good deal less of the massif was under forest than under grapes. From the high ridges with comprehensive views, the mountains looked like coalesced football stadiums, vineyards terraced into the sky. Above Palekhori, we picked and ate fourteen grapes, doubling the annual number destined for fresh consumption. At a table under a tree in Palekhori, we drank coffee that was good enough to eat. It was brewed in a briki scarcely larger than a gill. It came with a glass of cool water. "Palekhori" means old village. It is in the sheeted diabase, whose stately laminations are so distinct that you can all but see their lateral motion. Moores and Vine had found their best evidence of seafloor spreading at these altitudes. More recently, Moores had been studying fault blocks in the Troodos. In various tiltings of the sheeted diabase, he had seen that the rift valleys of ocean spreading centers tend to break into blocks as they widen—a version, on a small scale, of the faulting characteristic of the Connecticut Valley, for example, or the Newark Basin, or the Culpeper Basin, or the great western province of the Basin and Range.

The mid-ocean ridges run around the world very much like the stitching on a baseball, not in simple lines but in oscillating offset segments. Such a pattern evidently accommodates a sphere. In any case, it is what the earth looks like where it is pulling itself apart.

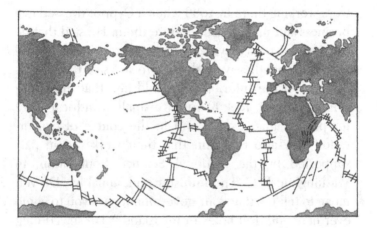

The ocean ridges of the world jump from rift valley to transform fault to rift valley to transform fault, everywhere they go. The rift valleys are typically about forty miles long, and are offset, also about forty miles, by the transform faults. Moores had found this pattern in the high Troodos. In rock beside a road, he pointed out sandstone sediments that fell into a bathymetric depression where a transform fault intersected a spreading center. A goatherd walked by—blue shirt, soft olive hat, a stick bag on his back. *"Yasas!"* he said, in greeting, above a din of goats. Nothing in his face suggested that he found it at all strange to come upon two men with the beards of Greek Orthodox priests squinting into coarse rock with the lenses of jewellers.

In a remote mountain valley, we walked into a box canyon and saw feeder dikes five metres wide that had driven upward through pillow lavas to break into the ocean and form more pillow lavas. Moores and Vine were working there in 1968 when a Cypriot dressed in a business suit materialized before them. He said that he was just out for a drive and had seen their vehicle, and he asked what they were doing. He also said that he was Minister of the Interior. They told him that they were drilling into the rock to remove small cylindrical cores for paleomagnetic data, and in the course of conversation they also told him that their work was in part supported by the National Science Foundation, in Washington, D.C. The minister said, amiably, "Do you mean to tell me that your government pays you to come over here and drill holes in my island?" He stayed long enough to learn that his island was the keystone of the sea. A year and a half later, he was assassinated.

In the way that the Smartville arc seems to have brought gold to California, Cyprus brought copper to the world. "Cyprus" means copper. Whether the island is named for the metal or the metal for the island is an etymology lost in time. The mining geologist George

Constantinou, director of the Cyprus Geological Survey, took us south from Nicosia one morning into the
hill country near a village called Sha. On soil eroded
from pillow basalt, he led the way into a grove of pines
that surrounded a pit forty feet deep. Like countless old
mines, it was partly filled with water. He said that the
pit had been there four thousand years. Constantinou
was a handsome man of alternately bright and brooding
aspect, with light wavy hair, strong features, and such
commanding stage presence that I imagined him as an
actor. I imagined him as Prince Hamlet, King Henry V,
and Archie Rice, because his physical resemblance to
Laurence Olivier was so close it was unnerving. Of this
small excavation framed by Aleppo pines he spoke with
resonance as well as reverence. Cypriots thirty-five centuries before Christ had walked into this pine grove,
and others like it, and had found native metallic copper
lying on the surface, he said. Pine resins in the groundwater had mixed with copper sulphate and reduced the
copper to metal.

When Cyprus was spreading in the Tethyan floor,
seawater descended through fissures and—close to or
within the magma—picked up quantities of dissolved
copper, and lesser amounts of mercury, manganese, tin,
silver, and gold. Like the black smokers active now in
the Red Sea and the Gulf of California, hot brine
plumes rose through the Cypriot rock and precipitated
metals and metallic compounds on the pillow lavas.
From everywhere in the ancient world, people turned
to Cyprus for weapons-grade copper. The swords,
spears, and shields of innumerable armies were made

from Cypriot copper. Before long, though, the resin-reduced metals were gone. More than a millennium passed before the Cypriots learned that dark earths where the metals had been were not a whole lot less cuprous than the metal itself. Rainwater—rare in Cyprus on the human scale but continual in geologic time—had removed lighter materials and had concentrated the copper minerals malachite and azurite in an upper zone of extreme high assay. Geologists of the twentieth century would describe such a concentration as a supergene enrichment. The ancients somehow discovered that if they mixed the cuprous earth with umber, and then heated the mixture, molten copper would flow. There was plenty of umber close at hand. Umber is an oxide of manganese and iron. In spreading environments on ocean floors today, umber is piling up on the pillow lava in large dark-chocolate cones beside the black smokers, as it did on the pillows of the nascent Cyprus.

In 2760 B.C., smelting began in Cyprus, Constantinou told us. And in the following centuries Cyprus became an island of seven kingdoms. Slag heaps developed in forty places. The Iliad is populated with warriors armed in bronze. Bronze is copper hardened by adding some tin, and the copper would have come from Cyprus. (Copper was mined on Cyprus for nearly two thousand years before the lifetime of Homer.) In 490 B.C., Darius the Persian attacked Greece with forty thousand soldiers who carried bronze shields and bronze javelins. The Phoenicians also mined copper on Cyprus, the Romans as well. The ancients stripped the supergene, and other rich ores, down to the water table, where they had to stop. The Republic of Cyprus once used ancient slag for roads, but the old slag heaps are now protected as monuments.

After the mine at Sha, we drove on ancient slag through roadcuts of pillow basalt, west-northwest. There were orchards of carobs, figs, and pistachios, and an understory of prickly pears. This was not Lawrence Durrell's north-coastal range of "silk, almonds, apricots, oranges, pomegranates, quince." It was an interior country of buff eucalyptus, of thousand-year-old olive trees fluted at the base. Nearly every farmhouse was white, and most had sky-blue shutters—the colors not of Cyprus but of Greece. On virtually all rooftops were boxy solar water heaters, like raised sarcophagi. With few exceptions, they carried advertising.

On the Mesaoria, we passed new, isolated towns— one-story clusters of temporary housing, marginally superior to internment camps, built to shelter Greek Cypriot refugees from communities north of the Attila Line. We left the highway and went into the new and extremely narrow streets of Peristerona, which seemed less a town than a military barracks. Many of its people were natives of Katokopia, scarcely three miles away, but Katokopia was lost to them, beyond the Turkish line. Among these was Anastasia Constantinou, the mother of the director of the Geological Survey. She was elderly, tall, dressed in black, obviously pleased to see her son no matter what he might bring with him. The hour was near noon, the air somewhat humid and above a hundred degrees. There was a small greenhouse full of gardenias, camellias, and azaleas. The geologist went among them with a mist propagator. He opened folding chairs on a concrete terrace with a view across the treeless plain to the marble mountains of Kyrenia—the dark wall of the forbidden north. His mother set on a table bowls of boiled rice flour that had lightly thickened as it cooled. Sprinkled with sugar, it was very cold, standing in rose water. In that volcanic heat, it had four times the effect of a cold fruit soup, twice the effect of gazpacho. If you closed your eyes, you saw pools among gardens descending into pools.

At Skouriotissa, southwest of Peristerona, the concurrence of geologic time and human time had been long enough to approach a record. A very large working strip mine there had been in operation for four thousand three hundred years. The slag, piled in pyramids, represented all that time. Constantinou said that there were at least two million tons of ancient slag in and around Skouriotissa. "Skouria" means slag. The massive copper-bearing sulphide ores of Cyprus have a very characteristic sugary structure, he said, resulting in an incompetent rock that was always easy to mine. "The ancients were excellent geologists. They knew the geology of the ore bodies of Cyprus. I am an exploration geologist with a Ph.D. I don't think we will find any ore body the ancients did not know about."

The earliest known smelting of copper was in China. Did the Cypriots figure it out for themselves or learn how to do it from others? In ancient manuscripts, Constantinou said, there is no insight that helps to answer that question. Where you find copper, you will find iron. The umber of Cyprus is more than half iron. "It seems logical to expect that Cypriot umber was the world's first iron source, that iron was invented on Cyprus. The ancients used gabbro to grind the ore. They lined their furnaces with serpentine."

They fired their furnaces with Aleppo pines, and other conifers—the ancient forests of Cyprus. To smelt one pound of copper from sulphide required three hundred pounds of charcoal. From the earliest beginnings of the mining until the last years of the Roman Empire, about two hundred thousand tons of copper were smelted on Cyprus. That used up fifty-eight thousand square miles of pinewood forest, on an island whose total area is thirty-six hundred square miles. The forest had to be rejuvenated sixteen times for copper alone, not to mention the fleets of ships that were made on Cyprus, or the firing of the island's world-renowned kilns.

"For lack of wood, Oman, Iran, Saudi Arabia, Egypt, and Israel all had short-lived mines," Constantinou said. "The Troodos gave water here to support trees. But sixty million tons of charcoal made from 1.2 billion cubic metres of wood is no joke. Sometimes I close my eyes and see that ancient scene. I get crazy. I see all those people, tens of thousands of people, carrying ore, carrying wood."

Before Moores went back to California and I to Switzerland, I accompanied him on a brief reconnaissance in Macedonia. While Cyprus was surely, as he had described it, one of the best-developed ophiolite complexes in the world, the sequence there did not include a large percentage of mantle rock. In a typical slice of ocean lithosphere, the mantle rock is nearly twenty per cent, bottoming at the aesthenosphere, the lubricious zone in the mantle which allows plates to glide. In Cyprus, the serpentinized mantle rock of the Troodos was a relatively small part of the peridotite that had once been included in the package. The rest was buried or lost. If you wanted to sense what was missing, you could do so in Macedonia, where seven vertical kilometres of exposed mantle constitute one of the thickest measurable sections of mantle rock emplaced on any continent. Moores said, "Presumably, it goes to the bottom of the plate."

In the disjunct cacophony of the airport in Athens—through a sea of Arabs wearing kaffiyehs and tobes and running shoes, of Ethiopians with wallets the size of magazines, of Panasonic briefcases turned up full—we found Hertz. We were soon at the foot of the Acropolis, establishing our bearings. Moores told me to notice the red shales and red cherts around the Theatre of Dionysus, on the low ground, and—over his shoulder as he rapidly climbed—to watch for the contact where the lithology changes. It would have been hard to miss. The freestanding, high-standing Acropolis is an almost pure massive limestone, sitting on the cherts and shales. Our way was blocked by a ticket booth. We paid a hundred drachmas. American college students were all over the summit. In the frugal shade of the Parthenon, Moores had the look and certainly the sound of a free-lance English-speaking guide. He mentioned Ictinus, Callicrates. American students leaned in to listen. He lauded the durable Phidias. He moved to the south side of the building, and students followed. He mentioned that limestone is soluble in water. Therefore, it includes caves. In caves within this hill, gods were thought to reside. Grottoed limestone will impound water. If you were seeking refuge, or a place to endure a siege, you would choose a hill like this one. We were looking south over the Stoa of Eumenes to the shore of the Saronic Gulf. From runways there, 747s were rising. They seemed in no hurry to go away. They seemed to hang like barrage balloons. Moores said, "After the battle of Salamis, ships beached themselves by the airport. As caves in limestone enlarge, their roofs eventually collapse."

He mentioned the Parthenon's historical stratigraphy—temple, church, mosque—and the erosional forces that had brought the building to its present condition: rain, acid rain, smog, gunpowder. In 1687, the Parthenon was in use as a Turkish powder magazine. Venetians bombarded it, and the powder exploded. The event was geomorphologically catastrophic. For two thousand years the Parthenon had stood there uneroded, until that night in 1687. "There is no mortar in the Parthenon," Moores added pensively. "It is all marble, and held together by gravity. And it's gone through earthquakes, too. The geology is not well worked out here. The general story is that the Acropolis is a klippe, resting on the red cherts and shales. It is not a deep block."

A klippe is a remnant of a nappe. A nappe is a large body of rock that has been moved—by gravity, by thrust-faulting, or by any other mechanism—some distance from its place of origin. If you liked, you could call Cyprus a special kind of nappe. Moores gestured to the east, across the white city and its numerous hills, to the serrated profile of the Hymettus Range, less than ten miles away. "It is thought that the Acropolis came from there," he said. "There are problems with the idea, but it is distinctly possible."

In the heat and pressure of a collision or some other tectonic event, limestone softens, recrystallizes, and hardens as marble. The Hymettus Range is for the most part marble. Its limestone first collected on the floor of Tethys, and was later folded in collisional mountains compressed by Africa moving northeast.

Marble quarries in the range had been there for something like three thousand years. The Parthenon came out of a quarry at the foot of Mt. Hymettus.

If the Acropolis is a klippe, the Acropolis itself came away from the Hymettus Range, in Eocene time, and travelled overland to Athens. About fifty million years later, in late Holocene time, the Parthenon followed, in carts.

"An alternative possibility," Moores said, "is that the Acropolis is a large block in a mélange with a matrix of red cherts and shales—an accretionary wedge tectonically scraped from Tethys when the seafloor was subducting in the Mesozoic."

The American students were looking at one another, and Moores was becoming self-conscious. A guide he may have been, but the language he was speaking was, to the students, local. "I have always thought it sacrilegious to come here and do geology," he said. The tone was apologetic but not sincere. His next words were "I think the shales would correlate with the Olonos-Pindos deepwater sediments, which extend from the Peloponnesus through western Greece."

As one might expect in a marine country sitting on a microplate caught in the crunch between Africa and Europe, ophiolitic fragments of varying age are strewn about Greece like amphora handles. As Moores drove up the broad avenue of Vasillisis Sofias and into Sintagma Square, he remarked that the silver that financed Athens came out of a metamorphosed ophiolite at Laurium, near Cape Sounion, the southern tip of Attica. He parked beside the Bank of Greece. Standing at a teller's cage on a green-and-black floor of polished serpentine, he changed dollars to drachmas.

North through Attica we moved swiftly, as if we were in a light plane flying at an altitude of three feet. A few miles west of the North Euboean Gulf, Moores pulled over and stopped at what appeared to be a long high roadcut. It was actually a limestone cliff, of the Parnassus Range, which rose steeply behind it. "There are bits and pieces of ophiolite on top of the Parnassus," Moores said. "This is Thermopylae." The broad coastal plain to the east was full of olives and cotton and was large enough to accommodate a very large army. In 480 B.C., however, the coastal plain was not there, and water lapped close to the rock. There was insufficient room for the attacking Persian Army. Leonidas, king of the Spartans, defended the narrow margin of land with Parnassus ridges at his back. He was defeated after the Persians learned of a route around the ridges. "Hot springs were at the foot of the mountains then," Moores said. "They are long gone. The coastal plain came up recently—at some point in the past twenty-five hundred years—as a result of an earthquake."

In the nineteen-sixties, in his long and lonely field seasons in Macedonia, Moores read his way in several senses into the country. He learned to speak the language. His interests were well spread across a couple of hundred million years. He asked me if I knew John Cuthbert Lawson's *Modern Greek Folklore and Ancient Greek Religion*, and described it as "polytheistic mysticism with a superficial patina of Christianity." Not being a geologist, I took his word for it. Farther north, he said:

——*This is Lamia. A few kilometres outside Lamia, during the Second World War, Greek partisans shot three German soldiers. Germans stopped the first hundred and thirty-eight people who came down the road, and killed them. The German rule was fifty Greeks for one German.*

——*Over there to the east, fifteen miles, is Volos, where Jason and the Argonauts sailed from.*

——*Pharsala is about twenty-five miles west of us. Near Pharsala, Caesar defeated Pompey on alluvium at the western margin of the Tsangli ophiolite. Pharsala itself is on serpentine. If you see a black church, you are probably looking at serpentine. That's particularly so in Italy. In Florence, the dark rock in the walls of the Duomo is serpentine—and the Giotto campanile and the baptistery with the Ghiberti doors. In Istanbul, the dark columns in the Hagia Sophia are serpentine.*

We came into a vast horizon of land so flat it seemed unnatural. It seemed flatter than the Great Central Valley of California, if that is possible. Solitary oaks were widely spaced above cotton, wheat, barley— a tree on each twenty acres, more or less. In a manner that called up ritual, tall telephone poles stalked across the two-dimensional landscape. Woodhenge. A huge Pleistocene lake had been here. In the stillness of its depths, smooth silts collected:

——*This is the Plain of Thessaly. In the eighth century B.C., Greek tribes settled here, chasing off the inhabitants. The fugitives went into the hills. They were fine horsemen. They raided the Greek settlements on horseback. The legend of the centaurs may have come from this. At night, you couldn't tell man from horse.*

——*Do you see those switchbacks climbing out of the plain? The Greeks used to survey a road by putting a hundred kilos on the back of a burro and sending him uphill. They followed the burro with a road.*

Three coastal mountains now formed the eastern skyline: Mt. Pelion, Mt. Ossa, and Mt. Olympus. In an island universe of Mt. Olympuses, there is one Mt. Olympus. This one. Sealed in its own integument, at the moment it was eighty per cent cloud. We could see only the base. Of the country more immediately around us, Moores said:

——*This is the Pelagonian Massif, a Mesozoic microcontinent that was thrust over a dome of younger sediment. The dome is marine rock—shallow-water limestone and flysch. From the highest part of the dome, the Pelagonian rock has worn away, and in that window stands Mt. Olympus, ten thousand six hundred feet. I think you need technical equipment to get to the summit.*

——*This is the Vale of Tempe, where the Muses came down off Olympus and played in the waters.*

We were soon in highland Macedonia, with wide views over red-and-white villages to far-distant mountains: west into the Pindus Range, north into Albania. At five thousand feet, the modest summits of Macedonia—Vourinos, Flambouron—were about as high as the Adirondacks of northern New York. The central Adirondacks could be said to be otherworldly, their lithology being rare on earth but nearly identical with most of the surface of the moon. In a different but analogous manner, Vourinos and Flambouron and the country roundabout related to surroundings far removed from the face of the earth. Somehow they had been lifted forty-five thousand vertical feet, and were almost pure mantle.

The Vourinos Complex—this mantle peridotite included—was Tethyan seafloor that formed in a spreading center in Jurassic time. It was emplaced on the Pelagonian microcontinent in lower Cretaceous time, and later broken into four major fault blocks. The narrative was straightforward and fairly simple, but of course it had not been helpful to Moores as he began work here in 1963, because plate tectonics and emplaced ophiolites were uncoined terms and the narrative did not exist. He was further inconvenienced by the ravages and deceptions of erosion, which had caused the lowest rock in the sequence to stand highest in the country. Moreover, the sheeted dikes of Vourinos diabase were misleadingly parallel to the sedimentary beds that had formed above them. After years of living with this terrane and taking it apart in his mind, Moores had come to realize that soon after the dikes formed at a spreading center they had been rotated ninety degrees from their original plane. In recent time, the four fault blocks had tilted over as well, and were like broken segments of a fallen ancient column. On paper, Moores had brought them into spatial coherence. He had worked here intensively for three years, and continually after that. He had recognized the contrast between the mantle rock and the magmatic rocks above it, and had become convinced that these parts taken together were in turn parts of a larger sequence.

Now, on the road to Skoumtsa, in the valley of a small clear stream, he was standing on the contact between the ophiolite and the Pelagonian rock on which it had been emplaced. Pelagonian limestone was overlain by serpentine that had been sheared up badly and turned into a messy schist as it skidded to a tectonic stop. The mantle peridotite had been serpentinized there where it scraped upon the continent, but as we moved away from that boundary the peridotite became purer to the point of zero serpentinization. The rock had never been magma. We were seeing (he presumed) the earth's mantle in an essentially unaltered state.

Since the ophiolitic column was lying on its side, we could advance overland on the face of the earth and lithologically descend deeper and deeper into the mantle. We did this on a very rocky and narrow road, some of which had been cut by engineering. After crossing the stream, we paused to look at a dynamited outcrop—a rough texture, dark green, knobby with pyroxene in a smooth matrix of olivine. "This is solid mantle," he said. "It's just about as fresh as you'll ever find, with the exception of mantle material in a diamond pipe. We are roughly five kilometres below the petrologic Moho, and that would have been about ten kilometres below the Tethyan ocean bottom, and fifteen kilometres below Tethyan sea level. Some people like to think that this rock slid by gravity into its position on the Pelagonian continental platform. From fifteen kilometres below sea level? How do you do *that* with a gravity slide?"

Now and again in that country we saw, in the peridotite, bands and lenses of dunite, which generally contains chromite, the source of chrome. The chromite looked like small, spattered blobs of tar, not like the grille of a Jaguar. The village of Khromion was ten kilometres away. During the Second World War, Germans used Greeks as forced labor there. Chrome was essential for the high-strength steel in the armor of tanks; and chrome, as it happened, sharpened the effect of armor-piercing shells. Peridotite with dunite in it is softened, more susceptible to erosion. "You can almost map the rock by looking at the relative roughness of the terrain," Moores said. "The smooth and grassy swales have dunite under them. The pure peridotite is in the rough ridges."

For a couple of years, Moores worked alone in Macedonia, attracting the attention of nothing much but mastiffs, which appeared out of nowhere. The mastiffs were protective of sheep, and hostile to geologists and wolves. In 1965, Moores turned up in the country with a wife (Judy had just graduated from Mount Holyoke), and later on with the Volkswagen bus and Geneva, Brian, and Kathryn. This attracted the attention of old Macedonian women, who lived on the Vourinos Ophiolite, and would appear out of nowhere. They admired the Moores children so much that they spat on them. This was a custom that warded off the evil eye. Judy learned to snatch baby Kathryn away from crones who adored her. In the Volkswagen one day, Judy turned around and saw Brian spitting on his teddy bear.

In Davis in 1989, when Kathryn had just turned sixteen she fell sick with acute mononucleosis, and her throat was so choked by swollen tonsils that she was completely unable to speak. The antidotal spit was recalled at the kitchen table. Kathryn wrote on a slip of paper: "Maybe it only lasts sixteen years."

It was in Macedonia that I asked Moores how he felt about being in a profession that had identified the olivine that people would be ripping the mountainsides to take away, and he said, "Schizophrenic. I grew up in a mining family . . . Now I am a member of the Sierra Club."

If you ask someone in Arizona where Crown King is, the usual answer is a shrug. Someone going home to Crown King would turn off the Flagstaff-Phoenix highway, raise a plume of dust on Bloody Basin Road, and go west-southwest toward mountains. The elevation of the basin is thirty-five hundred feet. The ridgeline ahead is seven thousand feet, and while you lurch and rattle toward it—as Moores and I did recently in a rented pickup equipped absurdly with cruise control— a good deal of time goes by but the mountains seem no closer. Moores said he remembered his father whistling a Schubert serenade on Bloody Basin Road. Moores was hearing it now, as he always had when crossing this particular stretch of country, although he had not been there in twenty years. In early slanting light, fields of prickly pears flashed like silver dollars.

We went through a one-house town. On the seat between us was an Exxon map of Arizona. "That was Cordes," I said. "Why is that place called Cordes?"

Moores said, "Because that was Bill Cordes' house."

The dust behind us thickened. We were now on the old Black Canyon Highway. When Moores was teen-aged, this unpaved and bridgeless thoroughfare lined with mesquite, cat's-claw, and paloverde was the main route from Prescott to Phoenix. We passed five buzzards eating a rabbit. While distance compiled, the mountains continued their retreat. After leaving Black Canyon Highway, we threw even more dust on saguaro cactus, agave, cholla, and ocotillo. We went through Cleator, a town that somehow managed to seem smaller than Cordes. There was a gas pump dating from the twenties which seemed to have died in the thirties. Beside it—in the open air—was a radio that had last heard the Blue Network. If you squinted hard enough into the cactus, you could see Dorothea Lange changing film. After more long miles, we began to climb, and now—in direct proportion to the gradient—the route evolved from an ordinary unpaved right-of-way into one of the oddest and certainly one of the costliest dirt roads in America. Its humble surface passed through roadcuts of exceptional engineering. It went through narrow defiles past vertical walls of competent granite blasted by construction crews in the century before. A railroad had once climbed three thousand feet there, its purpose being to help dismantle the mountains themselves, to ease down from Crown King in gondolas inexhaustible ores of hard-rock gold. "This railroad was an incredible feat of engineering, resulting in futility," Moores said. "The ore just wasn't there. Mine promoters are a breed apart. Their mentality is 'Of course it's there.' When I was ten, I heard a promoter say, 'We have a thousand tons of ore blocked out. When we get going, we're going to process a hundred tons a day.' He didn't stop to calculate that a hundred per cent of his ore would be gone in ten days. Mine promoters will believe anything, and so will their backers. The money came from New York, principally. The promoters were always looking for people with more money than smarts."

A voice said, "They looked in the right place."

Some of the deep cuts in the granite were cul-de-sacs, and the dirt road turned sharply before them. Trains had entered them, and then backed up across a switch and on toward the higher ground. Some cuts were low-sided, like the sunken lanes of England. In the course of the climb, the chaparral of the Sonoran Desert gave way to forest of ponderosa pine. The chaparral ran highest up south-facing slopes. We bypassed a tunnel, its ends now mostly caved in. When Moores and his three sisters were children, their grandfather convinced them that the tunnel was the home of a monster known as the Geehan. Long before they were born— after the mining scheme failed and the rails were removed—the Geehan moved in.

Crown King was at six thousand feet, in a subsummit swale—a few dozen buildings, spread through a mile of forest. When the air was still, the people could hear vehicles far down the switchbacks, climbing. They could tell from the rattles who was approaching. In the Crown King that Moores returned to now, not a great deal had changed. The same old welcome sign stood on the outskirts:

THE FIRE DANGER TODAY IS
EXTREME

The main street was a rocky swath of white granitic dust. Two pickups were parked by a retaining wall below the veranda of the general store. An apple tree thirty feet high grew out of the veranda. In a window, a red neon ring circled the word "Lite" under the painted words "U.S. Post Office." There was an old anonymous gas pump, still dispensing gas. It served the mountain. A white pole beside it flew the American flag.

Roads threading the declivities above Crown King led to various mines—gold mines, mainly, and silver and zinc—one of which served as the town well. The air and the forest were so dry, the community so high, that a ready source of water was beyond imagining. Yet water emerged from fissures at the Philadelphia Mine. In these exceptionally arid mountains, monsoonal rains arrive in August. Moores remembered an August day when four and a half inches fell in a single hour.

Crown King now had four telephone lines, four parties on each line. In the nineteen-forties, when

Moores was growing up, there was one line. When you turned the crank of one of the four phones in town, three other people picked up phones to hear your outgoing call. I asked him what else especially came back to him about those years. He said, "The smell of warm pine needles and the sound of the wind."

Under the ponderosas, on the dry needles, were granitic boulders. Playing hide-and-seek, he had hidden behind the boulders. When he and his friends played baseball, the bases were rocks. One year, there were six students in the Crown King school—grades 1 through 8. Usually, there were ten or fifteen. Moores climbed half a mile from home to school, sometimes in fairly deep snow. We went up there now, and found that a room was being added to the building, and thus it would become a two-room school. Academic privies had served when Moores was a child. They had been replaced by indoor plumbing. The wood stove was gone. In the ceiling, the chimney hole was covered with a board. The old classroom was otherwise the same, with its tongue-and-groove walls of horizontal boards, its long span of lead pipe supporting a curtain so that one end of the room could serve as a stage. There was a personal computer on the teacher's desk. The Lombard piano was an upright that Moores remembered. Playing a few bars on it, he found it "more or less in tune."

When Moores was a child, there was a piano in Crown King that belonged to his grandmother Annie Moores. She was from San Francisco, where, as a girl, she had routinely gone to the opera and returned home to play the scores from memory. After her husband became a miner and they began a life of moving from one remote mine to the next, her piano went with her.

It helped that he owned trucks. From the railroad at Flagstaff to the Colorado River he had trucked the steel of the Navajo Bridge. His name was Eldridge Moores, and he was primarily a small-scale pick-and-shovel all-around hard-rock miner whose body temperature became progressively higher in the presence of lead, zinc, copper, silver, and gold. In the middle nineteen-thirties, he was mining copper in the Verde River drainage when he decided to pick up the piano and haul it to Crown King. His son, Eldridge Moores, father of Eldridge Moores, worked for his father, Eldridge Moores. His son's wife, Geneva Moores, had a piano as well, and the two families in concert, in two Ford trucks, doremifasoled up the mountain.

Eldridge III—the future tectonicist, ophiolitologist, structural geologist, editor of *Geology*—was born in 1938. One of his earliest memories is of his father and grandfather saying that no one seriously engaged in mining would seek or follow the advice of a geologist. They said it often. Typically, Eldridge's father would say, "Huh—geologists. They think they can see through solid rock."

Five miles from Crown King and a thousand feet higher, Eldridge's grandfather tunnelled into solid rock. The family's Gladiator Mine was just below the ridgeline of the mountains—a ten-man operation that grossed about a million dollars in ten years. Gladiator was a gold mine with enough lead and zinc to be declared a strategic industry in the Second World War, so it was not shut down, as most gold mines were. The shafts and adits (tunnels) went into the mountain several hundred feet to the stopes—chambers with five-foot ceilings, angled with the gold vein at sixty degrees. They loosened the ore with pneumatic drills, and took out fifty thousand tons, getting half an ounce of gold per ton.

The main shaft was now covered with chicken wire and surrounded by rusting debris—a contusion in the mountain a century old. "You rely on the competence of the rock to keep the chambers open," Moores said. "You play it by ear. You develop a sense of what the good ore looks like. If it had a lot of flashing sulphides, it was ore." The sulphides were galena (lead) and sphalerite (zinc). "The lead and the zinc betrayed the gold. They were only two or three per cent of the rock, but they were the clue."

Soon after the war, Eldridge's grandfather moved on to something else, and his father, having found a place below the summit where the gold vein out-cropped, started a new adit there. The mine was called War Eagle. Modestly, it would support his family, with three hundred ounces a year. He went at it first with a pick and shovel, then with a hand-driven bull prick, and finally with a pneumatic drill, as the rock, ever farther from weather, became fresher and harder. In ten minutes, he could shovel a ton of rock, enough to fill an ore cart. "It was backbreaking work," Moores said. "But my father had a tough back." When we looked into the small, cavelike mouth of War Eagle, Moores said, "There's the vein. That little fault zone—that's what it is. I was here when he started the mine. The rusty streak in the rock marks the vein he was following. These old miners had a very good sense of where things would be in rocks. They would look at an outcrop, see a streak of iron oxide, and say, 'Ah, yes, this must be the vein.' They ranked geologists with garbagemen and dogcatchers. Most of the geologists they met were starving third-rate consulting geologists who came into small-mining areas looking for money. The tectonic and chemical models weren't in place yet, so the understanding of ore deposits wasn't very good. The miners had an intuitive feel for where things would be that was probably better than what a geologist coming in cold could give them. My dad sure thought that geologists were a worthless bunch, basically—people who came into the country to write reports for mining companies telling them what they wanted to hear."

Helping at War Eagle, Moores as a boy shoved the hand-propelled ore carts from the working face to the ore bin, outside the mine. He emptied the carts on the grizzly, an inclined steel grid. The smaller stuff fell through, into the bin. (Ore-vein rock, generally weaker than the rock around it, tended to break into small pieces.) The larger chunks rolled onto a steel platform. By hand, he sorted them, choosing what he thought was good ore, and heaving rejects aside. From the ore bin, the rock went down chutes into trucks. After the war, his father had acquired an International ten-wheeler with six-wheel drive, capable of hauling twelve tons. When Moores was learning to drive, his version of the family sedan with the automatic shift in the supermar-

ket parking lot on Sunday morning was an International
ten-wheeler between Crown King and the summit. The
mine's driveway, five miles long, had been engineered
and was maintained by his father. Not the least of its
features was a railless plunge on the outboard side. As
we inched along it in the cruise-control pickup, Moores
said, "This was my first driving experience, this road. In
that International, I was one scared little teen-ager." As
he practiced, he dragged a road grader behind the
truck. His father was back there, working the grader.

Sometimes he rode with his father to Phoenix in
the big truck, on a dirt road flanked by desert, first en-
countering pavement in country that is now city, seven
miles north of the state capitol. The hot truck stank of
transmission oil. When the truck quit, they revived it.
Sometimes they would be stranded for hours by a flash
flood.

The family lived for some years in the most impos-
ing house in Crown King, which they rented for
twenty-five dollars a month. It is squarish, board-and-
batten, on a rocky platform behind a retaining wall. A
refrigerator reposed on the front porch. Moores called
that "an emblem of rural Arizona." When he was nine,
his family bought another board-and-batten house, with
a yard of dry needles and granite boulders. It was light
blue now, with white trim and a tin roof, and suggested,
in its setting, the quarters of a forest ranger. There
were two bedrooms and three daughters, so his father
brought a shack from the mine and set it on a poured
slab, six feet by nine, and that was his son's freestanding
bedroom. The main house was heated by unvented bu-

tane. The well went dry for several months each year. Eldridge's father put a tank of water in a dump truck, and raised the body. The family had its own water tower.

At the age of ten or eleven, Eldridge noticed with some interest that two of the large rocks close to the house were different in color, yet each was called granite. He was somewhat puzzled, but his hair did not stand on end. This was the one touch of geological curiosity that he felt throughout his youth in the mining camp. In his portable bedroom, there was nothing that even vaguely resembled a mineral collection. Where a budding herpetologist might have a closet full of snakes, a chemist a set of volatile powders, a cosmologist a wheel of stars, Eldridge had musical instruments. When his father opened War Eagle, Eldridge's interest was keen: "I wanted them to find the good stuff, because it put shoes on the feet. But I wasn't curious about the vein."

Any rock that was hard and dark was blue diorite to the miners; anything platy was schist; everything else was granite. One did not have to go to Caltech to learn this geology. Endlessly, his father and other miners talked about the provender of rock. They sat on their porches in front of the refrigerators and reminisced about mining camps, mining failures, and yields in ounces per ton of ore. Eldridge's mind was elsewhere. Even before he entered his teens, he dreamed of places far from the ridge. He would forever remember Carl Vanlaningham, a friend of his father, remarking one day, with a glance around town, "Optimism is highest at the beginning. A mining camp has nowhere to go but down." Eldridge as a child had sensed this in a general and pervasive way. One day when he was accompanying his parents from one switchback to the next on the interminable road to Crown King, he suddenly burst out, "I've had it! If I never do another thing, I'm going to go out of here and stay out of here." His parents looked sad. He was ten years old.

He finished eighth grade and enrolled at North Phoenix High School when he was twelve. His father had built a house on the outskirts of Phoenix to accommodate his children's education. Eldridge's mother stayed with them. After she became the teacher at the Crown King school, his sister Carolyn (two years older than he) was in charge of the household in Phoenix. Moores' developing opinion of developing Phoenix was a good deal lower than his opinion of Crown King. As he would explain in later years, "there's something wrong with a place that looks to Miami for its cultural leadership."

In high school, his principal interests were music and history. Teachers urged the cello on him, because his hands were large. He has nearly perfect pitch. "If you ask me to sing a note, I can get within a few cycles per second. If I'm listening to a piece on the radio, I can tell you what key it's being played in." In Crown King, while his father struck notes on the family piano Eldridge with his back turned could identify the notes. Eldridge has said of his father, "Music meant a lot to him, too. But, as he saw things, men didn't go into music. They had to go into something that was practical."

To fulfill the high school's requirements, Eldridge also studied science and math. He was not inwardly driven toward a scientific or technological career, but he was already being nudged in that direction by what he has described as "a regional force." He explains, "In that part of the United States, it was assumed that any bright high-school student would go into science or engineering." His grade-point average was almost unimprovable. North Phoenix High School took particular pride in steering its best students toward the California Institute of Technology. Caltech offered him a larger scholarship than any other school to which he applied. Tractably, compliantly—sixteen years old—he went to Caltech.

Academically, he was at home there. For him, Caltech proved to be no more formidable than the one-room school in Crown King. Eventually, however, the day arrived when he had to choose a major—to decide, in effect, what he wanted to do with his life—and he experienced a sort of intellectual ambush. To his considerable surprise, he came to realize that there was only one discipline at Caltech that appealed to him strongly enough for such a commitment, and the discipline was geology. Under all his early indifferent attitudes, not to mention his avowals to escape from Crown King, there had obviously lain ambivalence. Evidently, what he thought he hated he did not altogether hate. He wondered still about those colors in granite. He may not have cared how the gold got out of the mountains, but he did want to know how the mountains came there to receive the gold. He remembers looking out through classroom windows, seeing the San Gabriel Mountains, and wishing he were up there. In the years ahead of him, he decided, he would like to combine history with science, and to travel the world out-of-doors. Those prerequisites could be combined, and restated as a single word. So he majored in geology. Today, if you look at him closely through your hand lens and ask him why he did that, he will give a little shrug and say, "I grew up in the mountains."

He will also say, "I had a hard time coming to grips with going into geology. While I was in graduate school, I still wondered if I could make a career in music."

At the mouth of the Gladiator Mine, we set up a bench with a discarded plank, resting it on discarded ores. Not many feet away were irises planted by his grandmother Annie Moores, whose house had once stood beside them. Just below the ridgeline, looking east, we ate sandwiches and spread on the ground before us the geologic map of Arizona. Six million acres of the original were also spread before us. We could see the Superstition Mountains, east of Phoenix. We could see the Four Peaks of Mazatzal Land. We could see a hundred miles. I remembered him once shaking his head with amazement at the joy experienced by a paleontologist who, in ten hours bent over in a blistering Wyoming gully, might find a couple of sharks' teeth. "I'm a ridge man, not a ravine man," Moores remarked at the time. "I like to get up and look out." Now in the foreground three thousand feet below us was the valley we had crossed in coming to Crown King. The axis of the valley, he said, running his finger along a black line on the map, was the Shylock Fault—"a major zone of tectonism that is reminiscent of major mélange zones that characterize consuming plates." Like the rock of the mine, the rock of the valley was Precambrian. He had found there—in sequence—serpentines, gabbros, and basalts. Pillow basalts. All this had suggested to him "plate activity eighteen hundred million to two thousand million years ago"—a collision, a docking, an addition to the continent arriving. He had published an abstract on the subject. Precambrian tectonics are, in their great antiquity, extremely difficult to read, but the rock of the valley suggested to him that he grew up in exotic terrane.

The granite of Crown King, he said, was technically quartz monzonite—a granite sibling, almost a twin. Where it contained a little iron, it would be pink. It had come into the earth molten, as a pluton, about seventeen hundred and fifty million years ago, and the country rock it had intruded was the rock here at the mine. (We had crossed the contact driving up from Crown King.) The rock at the mine was a rusty-looking, darkish, metavolcanic metasediment, two thousand million years old. Kicking at it, Moores said, "I would rank these rocks as not particularly easy to work with."

I thought he was referring to the picks, the shovels, the pneumatic drills—the backbreaking labors of his father and his grandfather. But he meant the geology.

"It's taken me a long time to get this stuff out of my system," he continued. "Metal deposits are telling you something interesting about tectonic systems." Thus, by the front door, his interests returned to the place where he developed. The livelihood of his family had depended on the yield of this rock. The relationship of an Arizona pluton to the rock in which it had ballooned as magma was now heating his imagination, as, before, it had heated his grandfather's and his father's.

They died in 1949 and 1979. "I don't have gold fever," Eldridge said. "When I get it, I stamp it out. I avoid the study of ore deposits, except as they are a scientific subject. Small mining is dirty, dangerous, boring, and dismal. After a while, watching people do it gets to you. The prospect of doing that made me want to get out, get away. I developed a hankering for places that were dust-free. I promised myself I was going to live in a place that was green and cool. Basically, this kind of mining is a futile operation. No one ever gets rich. It's got to be something in your blood."

I said, "Wouldn't you say it's in *your* blood?"

He said, "Sort of like an antibody."

From the Auburn suture of the Smartville Block, where you glimpse for the first time (westbound) the Great Central Valley of California, the immense flatland runs so far off the curve of the earth that its western horizon makes a simple line to the extremes of peripheral vision. In California's exceptional topography—with its crowd-gathering glacial excavations, its High Sierran hanging wall, its itinerant Salinian coast—nothing seems more singular to me than the Great Central Valley. It is far more planar than the plainest of the plains. With respect to its surroundings, it arrived first. At its edges are mountains that were set up around it like portable screens.

While looking out on the Great Central Valley one time, Moores described a winter day when he took off from Newfoundland in a snowstorm and flew to Toronto, landed there in strafing sleet, and flew on to Chicago (horizontal sleet), and then on across the high blizzarded plains and the Rockies (whited out) and the snowed-under Basin and Range to descend over the snowpacked Sierra, and bank toward a landing in a world that was apple green. The Great Central Valley was cool and moist and not cold and frozen, and for him, now, it was home.

It's not always cool, or—heaven knows—moist. Trunks of fruit and nut trees are painted white to keep them from being sunburned. Summer days commonly rise above a hundred, but the air will fall toward the fifties at night.

The ground surface is so nearly level that you have no sense of contour. A former lakebed can be much the same, where sediments laid in still water have become a valley floor. Such valleys tend to be intimate, however, while this one is fifty miles wide and four hundred miles long. It is not a former lake, although in large part it is a former swamp. Geology characteristically repeats itself around the world and down through time, but—with the possible exceptions of the Chilean Longitudinal Valley and the Dalbandin Trough in Pakistan—the Great Central Valley of California has no counterpart on this planet.

Engineers designing roads in the valley are frustrated by the lack of topographical relief. They have nothing to cut when they're in need of fill. If a new highway must cross over something, like a railroad track, the road builders go back half a mile or so and sink the highway into the earth in order to dig out enough dirt to build ramps to a bridge jumping the railroad. From the middle of this earthen sea, the flanking mountains are so low and distant that the slightest haze will give you the feeling that you are out of sight of land.

Over open ocean, the number of miles you can see before your line of sight goes off the curve of the earth is roughly equal to the square root of your eye level in feet. If your eyes are forty-nine feet up, you can see seven miles to sky. The formula is of very little use on land but is practical in this valley. When Darwin, off the Beagle, was travelling in Argentina, he sensed the subtle contours of the pampas:

For many leagues north and south of San Nicolas and Rozario, the country is really level. Scarcely anything which travellers have written about its extreme flatness, can be considered as exaggeration. Yet I could never find a spot where, by slowly turning round, objects were not seen at greater distances in some directions than in others; and this manifestly proves inequality in the plain.

He could have done the same in the Great Central Valley. Much of it is near sea level, but it does rise. North from Interstate 80, the valley rises steadily, with a grade so imperceptible it is measured by laser. Moores compares this rise to the inclined sides of a mid-ocean ridge. With respect to the abyssal plains, the ridges rise about six thousand feet, but over so great a distance that, in his words, "if you were put down on a mid-ocean ridge flank and told to walk toward the ridge crest you would not know which way to go." The grade of a fast-spreading ocean ridge, like the East Pacific Rise, is about the same as the grade of the Great Central Valley between Sacramento and Redding.

You watch magpies over the valley, larcenous even

in flight. You watch crop dusters—buzzing up and
down, up and down, like trapped houseflies. Rice is
sown from the air. Where DDT was once laid down
in an aerosol, fish—as living insecticides—are now
dropped from airplanes. They live in the checks, as
paddies are called, and eat larvae. In this essential flatness, where there is no visible relief, instruments can
find a minuscule astonishing topography; as if on a
smooth board very slightly weathered, the unevenness
of the land is discovered. Your eye can't discern it, but if
you were a rice rancher you would be dealing with it
every day. Rice seedlings need to stand in enough water
to cover them but not enough to kill them, and this del-
icate margin has caused rice fields—from the air—to
reveal the structure of the valley.

When the rice was first planted, vehicles bearing
surveyors' rods drove in circles around fixed transits to
discover the valley's contour lines, and along them

berms were nudged up a few inches to contain the shallow water. Sinuously paralleling the contours, the berms made the rice fields look, from the air, like supermagnified topographic maps. They showed tectonism expressible in centimetres. They showed the noses of anticlines, the troughs of synclines, microfolds, depressions—all too minimal to be detected by the human eye. Even very shallow water would run off these surfaces in every direction. So the rice plantations were terraced—each check, in altitude, scantly different from the next. Just as volcanologists use lasers to sense expansion in eruptive ground, rice ranchers have in recent years surveyed their rice fields with rotating beams. Contours are adjusted by the laser-controlled blades of earthmoving triplane levellers. The newer berms tend to be straight and less indicative of the geologic structures beneath them. There are five hundred thousand acres of rice in California. The climate is much the same as the climate in Egypt, which has the highest-yielding acreage of any rice country in the world. In the office of Jim Hill, who teaches rice at Davis, a sign says "Have a Rice Day."

The Great Central Valley is drained by two principal rivers, one flowing south and the other north. They meet in the valley and discharge themselves together into San Francisco Bay. The north-flowing river is the San Joaquin. The south-flowing river is the Sacramento, with its tributary the Feather River, which is dammed to reserve the snowmelt of the Sierra Nevada, not only to flood the rice fields and irrigate the other crops of the valley but also to travel six hundred miles in a life-support tube that is taped to the nose of Los Angeles. Rivers with common deltas are rare in the world. It would be difficult to name more pairs of them than the Kennebec and the Androscoggin, the Ganges and the Brahmaputra, the Tigris and the Euphrates, the Sacramento and the San Joaquin.

The floodplains of the Sacramento and the San Joaquin are dozens of miles wide. Before they were drained, and checkered like kitchen floors in shades of patterned green, they were known as the tulares or the tules or the tule swamps, in honor of the tough bulrushes that (pretty much alone) were able to survive the scouring inundations of spring. Sacramento stands only a little higher than the tules. To the conventional wisdom that one ought never to build on a floodplain, California has responded with its capital city. Old houses of Sacramento are in a sense upside down. Long exterior stairways lead to porches and entrance halls on upper floors. Whole neighborhoods are on stilts.

Emerging from the Sierra foothills among the blue oaks of the grass woodland, Interstate 80 rests on gravelly loams until it reaches the silty clays and humic gleys of Sacramento. Now I-80 has become a long elevated causeway that reaches across rice and sugar beets, marsh grass and milo, as if it were in search of Key West instead of San Francisco. In years of exceptional floods, fifty million acre-feet of water have come under the causeway. Some row crops and tree crops are on the deeper lighter soils near the river—the natural levees, where floodwaters give up the larger part of the material they carry. Eventually, imperceptibly, the ground beyond the floodplain goes up a few inches onto the outer lenses of an alluvial fan, a fine loam that has spilled eastward from the Coast Ranges like thin paint. The difference it makes is widely expressed in field crops, truck crops, orchards. In soil taxonomy, there are ten groups in the world. Nine are in this valley. Each is suited to a differing roster of crops. Plums, kiwis, apricots, oranges, olives, nectarines—it is the North American fruit forest. In some parts of the valley, roots are inhibited by a zone of hardpan. The soil's B-horizon, firmly cemented with silica, is like a concrete floor. Farmers used to bore holes in it and grow trees in the holes. Now they use diesel-powered earth rippers. Especially, this is the valley of beets and peaches, grapes and walnuts, almonds and cantaloupes, prunes and tomatoes. That is to say, the Great Central Valley of California grows more of each of those things than are grown in any other state of the union.

In 1905, the College of Agriculture of the University of California, in Berkeley, set up an experimental farm in Davis, Yolo County, in the valley's center. In 1925, the farm itself became an agricultural college. In 1959, it became a general campus in the state's university system. The livestock-judging pavilion is now a Shakespearean theatre. Under skyscraping water towers, the ground-hugging university is of such breadth and grandeur that it has its own beltway. It may have more bicycles than Shanghai. But Davis is still the main agricultural research center in California, and just outside the glassy postmodern geology building are sties containing massive monolithic pigs.

From the geology-building roof, Moores has looked across the immense flat ground and picked out features of Yosemite, a hundred miles to the southeast. Fifty miles southwest, he sees the detached oddness of Mt. Diablo, protruding beside the Coast Ranges. Looking the other way, he has seen Lassen—a white cone on the northern horizon, a hundred and forty miles distant—and, on nearer ground, the Sutter Buttes, a recent volcanic extrusion that has left a ring of jagged hills standing in the valley like a coronet set on a table. These landmarks embrace more than six million acres, or—in one look around—a landscape larger than Massachusetts. It is less than half of the valley.

When Moores arrived, in the nineteen-sixties, ten thousand people lived in Davis. Although the number has increased fivefold, Davis is still a quiet town, still a field station. From its shaded streets, crops and orchards reach out in all directions. To an open-air market in Davis on Saturdays farmers bring their lime thyme, their elephant-heart plums, their lemon cucumbers, bitter melons, and peaches the size of grapefruit. They bring tomatoes that are larger than the peaches, and tomatoes of every possible size down to tomatoes the size of pearls. Yolo County grows about as many tomatoes as Florida does. Yolo County grows ketchup in the form of "processing" tomatoes that could sit on a tee and be driven two hundred yards.

The Mooreses' turn-of-the-century farmhouse is "in the county," down a long thoroughfare of black walnuts at the edge of town. Their street is named Patwin, for the tribe that preceded the farms and the walnuts.

The house faces north, across tomato fields. From the east windows you can sometimes see the low line of the Sierra, and from the west windows the Coast Ranges, but there is no sense of valley. The word seems misapplied. As the edges of a flat so vast, those montane curbs fail to suggest the V that a valley brings to mind.

When Moores looks out upon landscapes, he sees beneath them other landscapes. Like most geologists, he carries in his head a portfolio of ancient scenes, worlds overprinting previous worlds. He sees tundra in Ohio, dense forestation on New Mexican mesas, the Persian Gulf in the Painted Desert. Once, after a day in the Sierra, while he was sitting outside at home beneath what appeared to be a chandelier of apricots, I asked him to describe this one particle of the planet, his own backyard, at differing times. He responded in the present tense, as geologists often do, while his narrative went backward, scene by scene—episodically, stratigraphically—disassembling and dissolving California.

In the late-middle Pleistocene, when pulses of alpine ice are appearing on the Sierra, this place would be much the same, on the natural levee of a creek, surrounded not by fruit trees but by swamps. Valley oaks are on the dry ground, inches above the swamps. The mountains—east and west—are even lower. Mt. Diablo is not there. It has "scarcely begun to grow."

Three million years before the present—in the Piacenzian age of late Pliocene time—neither the Coast Ranges nor the Sierra is above the horizon. From the hinge under the Great Central Valley, the Sierra fault block begins to rise. The tectonic behavior of the

Coast Ranges is different. Sluggishly, they come up from the deep. They have no integral structure. They are a fragmentary mass, a marine clutter. They will be known in geology as the Franciscan mélange. Appearing first as islands, the Franciscan pushes against the level sediments of the coastal plain and bends them upward until they are nearly vertical. Up through the mélange come volcanoes that spew lava and tuffaceous ash in and around the Napa Valley. There are active volcanoes on the crest of the Sierra. And between the nascent ranges the Sutter Buttes erupt.

I have a question. Why all this Fourth of July geology as recently as three million years ago, when all we have in these latitudes now are run-of-the-mill earthquakes?

Because in the Pliocene a triple junction of lithospheric plates is just off San Francisco, Moores replies. A subduction zone is dying out as its trench turns into the San Andreas Fault. The volcanism relates to that. For tens of millions of years, a lithospheric plate of considerable size lay between North America and the Pacific Plate. It is known in geology as the Farallon Plate. By the late Pliocene, this great segment of crust and mantle, possibly at one time a tenth of the shell of the earth, had in large part been consumed. Fragments of it remain: in the north, the Gorda Plate and the Juan de Fuca Plate, whose subduction under North America has produced Lassen Peak, Mt. Shasta, Mt. Rainier, Mt. St. Helens, Glacier Peak, and the rest of the volcanoes of the Cascades; in the south, the Cocos Plate and the Nazca Plate, whose subduction has created Central America and elevated the Andes. For tens of millions of years, the Farallon Plate went under the western margin of North America, while North America gradually scraped off the Franciscan mélange of coast-range California. To the west, under the ocean, was the spreading center that divided the Farallon Plate from the Pacific Plate. As the Farallon Plate, moving eastward, was consumed, the spreading center came ever closer to California. At Los Angeles and Santa Barbara, the Pacific Plate first touched North America, twenty-nine million years before the present. Where it touched, the trench ceased to function, the spreading center ceased to function, and the plate boundary became a transform fault. It was only a few miles long at first, but steadily the great fault propagated from Los Angeles and Santa Barbara to the north and to the south, shutting the trench like a closing zipper. The triple junction of the Farallon Plate, the Pacific Plate, and the North American Plate migrated northward with the northern end of the fault. And so, in the Pliocene, three million years ago, the triple junction was off San Francisco. The volcanoes in the Sierra were the dying embers of Farallon subduction. The volcanoes in the Napa Valley and adjacent coastal ranges were a result of the new fault pulling the earth apart at kinks and bends. The eruption of the Sutter Buttes almost surely relates to the dying subduction or the new plate motions but is, as they say, not well understood. Now, in the Holocene, the triple junction is still moving north. For the moment, it is at Cape Mendocino, where the San Andreas ends and what is left of the Farallon Trench continues. That is how things appear, anyway, in present theory.

Six million years before the present, in the late
Miocene, Moores and his apricot tree would be in or
beside a saltwater bay that covers most of the Great
Central Valley. It is full of tuna and other large fish, because an upwelling of cold water (like the upwelling in
the Humboldt Current off modern Peru) has filled the
bay with nutrients. There is no Golden Gate. The bay's
outlet is at Monterey. A terrane is moving along the
west side of the San Andreas Fault. Carrying with it
the sites of San Diego, Los Angeles, Santa Barbara, San
Luis Obispo, Big Sur, Monterey, and Salinas, it will
someday be known as Salinia.

In the Eocene, fifty million years ago, Moores'
backyard in Davis is mud at the bottom of the Farallon
Ocean, some thirty miles offshore, on the continental
shelf. As Eocene rivers pour into these waters—having
advanced their gravels from Tibet-like altitudes and
across the low country that will one day rise as the
Sierra—they cut submarine canyons through the future
Great Valley. The rock that preserves this story is a
marine shale, loaded with shelf creatures of Eocene
age. Below Davis, in the Great Valley Sequence of sediments, it lies about twenty-five hundred feet down.

In the Cretaceous, some eighty or ninety million
years ago, Moores' address is a precariously inclined
deep-sea fan—a spilling of sediment down the continental slope toward the trench where the Farallon
Plate is disappearing. About sixty miles wide, the trench
lies in the space that will one day separate San Francisco and Fairfield. As the slab of the Farallon Plate
melts beneath North America, it contributes to the

magmas of the great batholith and the superjacent volcanoes of the ancestral Sierra Nevada.

At the end of the Jurassic, about twenty million
years after the docking of the Smartville Block, another
island arc comes in and docks against Smartville, more
or less directly under Moores and his tree. Geology will
call it the Coast Range Ophiolite, and it will lie under
forty thousand feet of Great Valley sediments and be
warped into the coastal mountains. One of its large
fragments will end up in the Oakland hills.

When Smartville docks, in the Jurassic, its individual islands possibly resemble Hokkaido, Kyushu, and
Honshu. The trench closes east of Sacramento and a
new one opens west of Davis and begins to consume
the Farallon Plate. The downgoing slab of the Farallon
Plate depresses the region and creates the structural
basin that will fill up with sediments and become the
Great Central Valley. Since there will be no Sierra
Nevada and no Coast Ranges for nearly a hundred and
fifty million years to follow, the result will be a valley
that is not a conventional river valley but a structural
basin filled to the brim with sediments that (almost
wholly) do not derive from the mountains around it.

Before Smartville, blue ocean—extracontinental,
abyssal ocean. In the earliest Triassic, the site of Davis
is far out to sea. The continental shelf is back in Idaho
and Nevada. North America in these latitudes has been
growing. Two terranes have already come in. But here
at the dawn of the Mesozoic the continent has not yet
received so much as a hint of California.

The Napa Valley is thirty-five miles due west of Davis—an easy run for a field trip, a third of it flat and straight. The occasions have been several, not to mention spontaneous, when Moores and I have made westering traverses, collecting roadside samples of rock and wine.

After the level miles of field crops and fruit trees and almond groves, the ground suddenly and steeply rises in oak-woodland hills, so brown and dry for much of the year that geologists working among them can accidentally start fires with sparks from their hammers. Putah Creek, the stream that has spread its fine silts to Davis, is here a kind of door to the Coast Ranges, spilling forth their contents, coarsely bedded. Among the stream's cutbank gravels are layers of air-fall tuff that descended from the coast-range volcanoes of the Pliocene, and conglomerates that contain serpentine pebbles, peridotite pebbles, chert pebbles, graywacke pebbles, volcanic pebbles—the amassed detritus of several geologies, suggesting the commotion in the rock to come. Also present are fine-grained remnants of extremely fluid basalts that burst out in the northwest in middle Miocene time, covered areas the size of Iceland in a single day, and are thought to have been the beginnings of the geophysical hot spot that has since migrated to Yellowstone. The Columbia River flood basalts, as they are known, reached their southern extremity here.

As we go up the stream valley and arrive at the shore of Lake Berryessa, we pass through huge roadcuts of sedimentary rock whose bedding planes, originally horizontal, have been bent almost ninety degrees and are nearly vertical. Reaching for the sky in distinct unrumpled stripes, the rock ends in hogbacks, jagged ridges. Cretaceous in age, these are the bottom layers of the Great Valley Sequence, bent high enough to resemble the bleaching ribs of a shipwreck. They are some of the strata that were folded against the Franciscan mélange when it rose (or was pushed) to the surface as the latest addition to the western end of the continent. In the heat and pressure of the Farallon Trench, the Franciscan sediments had been metamorphosed to varying extents, with the result that when they ultimately appeared on the surface they were miscellaneous and heterogeneous well beyond the brink of chaos. This lithic compote is the essence of the Coast Ranges. You leave the precise bedding planes and jagged ridgelines of the Great Valley Sequence and enter a country of precipitous nobs and rootless outcrops resting in scaly clay. In its lumpiness it resembles a glacial topography magnified many times. If the Great Valley Sequence can be compared to regimental stripes, the Franciscan is paisley.

"Look at this munged-up Franciscan glop!" Moores exclaims.

Narrow thoroughfares twist among the giddy hills. Ink Grade Road. Dollarhide Road.

"Look at that mélange! Holy moly, look at the lumps!"

Between the grinding lithospheric plates, the rock of this terrain was so pervasively sheared that a roadcut in metabasalt looks like green hamburger. We clearly see its contact with the scaly clay.

"That clay is the matrix of the Franciscan, in which blobs of various material are everywhere contained, and that is the guts of the Coast Range story. The metabasalt is a tectonic block in the matrix. You can see why people who tried to map stratigraphy went crazy. Imagine—before plate tectonics—the aching problems that this fruitcake, this raisins-in-a-pudding kind of stuff, produced. It doesn't fit the stratigraphic rules we all grew up on. It was assumed that you had a stratigraphic sequence here, and for years people tried unsuccessfully to explain these places in terms of eroded and deformed stratigraphies. In 1965, Ken Hsü proposed the mélange idea. But he suggested that the mélange had come here by gravity—that it had slid off the Sierra. No one had the idea of underthrusting—

what we now see as the subduction of one plate beneath another, with all this miscellaneous material
being scraped together and otherwise accumulating
at the edge of the overriding plate. In 1969, Warren
Hamilton, of the U.S.G.S., published a paper on the
underflow of the Pacific crust beneath North America
in Cretaceous and Cenozoic time. He presented the
paper at the Penrose conference on the new global
tectonics. Suddenly, people had a new view of the Franciscan. They said, 'Oh, that must be a berm resulting
from subduction.' And the whole story broke open."

The Franciscan mélange contains rock of such
widespread provenance that it is quite literally a collection from the entire Pacific basin, or even half of the
surface of the planet. As fossils and paleomagnetism indicate, there are sediments from continents (sandstones
and so forth) and rocks from scattered marine sources
(cherts, graywackes, serpentines, gabbros, pillow lavas,
and other volcanics) assembled at random in the matrix
clay. Caught between the plates in the subduction,
many of these things were taken down sixty-five thousand to a hundred thousand feet and spit back up as
blue schist. This dense, heavy blue-gray rock, characteristic of subduction zones wherever found, is raspber-
ried with garnets.

In a 1973 paper by Kenneth Jingwha Hsü appears
a sentence describing the Franciscan mélange—this
five-hundred-mile formation, the structural nature of
which he was the first to recognize—in terms that could
be applied to almost any extended family sitting down
to a Thanksgiving dinner:

These Mesozoic rocks are characterized by a general destruction of original junctions, whether igneous structures or
sedimentary bedding, and by the shearing down of the more
ductile material until it functions as a matrix in which fragments of the more brittle rocks float as isolated lenticles or
boudins.

Hsü was born in China and began to use his um-
laut as a tenured professor at the Swiss Federal Institute of Technology.

The mélange above Auburn, which collected
against North America before the arrival of the Smartville Block, tells the same sort of story as the Franciscan, with the difference that the rock in the Sierra
mélange has been almost wholly recrystallized, as a result of the collisions that completed California. Kodiak
Island and the Shumagin Islands are accretionary
wedges, too—shoved against Alaska by the northbound Pacific Plate. The Oregon coast is an accretionary wedge (the Juan de Fuca Plate versus the North
American Plate), complicated by a chain of seamounts
that have come drifting in, making, among other things,
Oregon's spectacular sea stacks. The outer islands of
Indonesia are accretionary berms like the California
Coast Ranges (the Australian Plate versus the Eurasian
Plate), not to mention the Apennines of Italy, the north
coast of the Gulf of Oman, and the Arakan ranges of
Burma.

Now and again in the Coast Ranges you see ophiolite pillows on top of the mélange—a typical relationship, since the mélange forms at the edge of the

overriding plate and the ophiolite is already on the overriding plate, having been previously emplaced there. Ocean-crustal detritus is widespread and prominent among the rocks of the Franciscan, but the Coast Range Ophiolite, in more concentrated form, is in the eastern part of the mountains, where it has been bent upward with the overlying Great Valley sediments, and pretty much shattered. Between Davis and Rutherford is a block of serpentine—disjunct, floating in the Franciscan—that underlies the bowl of a small mountain valley. The serpentine has weathered into soil, now planted to vines. These are some of the few grapes in California that are grown in the soil of the state rock. Moores is predisposed toward the wine. To him, its bouquet is ophiolitic, its aftertaste slow to part with serpentine's lingering mystery. To me, it tastes less of the deep ocean than of low tide. The stuff is fermented peridotite—a Mohorivičić red with the lustre of chromium.

The winery is in the deep shade of redwoods on a tertiary road. It makes only ten thousand gallons and has been in one family for a hundred years. The cave is in Franciscan sandstone. The kegs, tanks, and barrels are wood. Outside the cave, we stand on a wooden deck looking into a steep valley through the trunks of the big trees. Passing a glass under his nose, Moores remarks that the aroma is profound and reminds him of the wines of Cyprus. There is an intact ophiolite on the side of Mt. St. Helena, at the northwest end of the Napa Valley, he tells me—an almost complete sequence, capped with sediments but lacking pillows.

There's a complete sequence on the east side of Mt. Diablo. "If you mapped the Coast Range Ophiolite, it would go from Oregon all the way down, in discontinuous blobs, plus the shards you see around San Francisco and elsewhere—rocks of the ophiolitic suite that just lie around as broken pieces, like the block that is under these grapes, and cannot be read in sequence." When Moores was first in California, he happened upon a report about mercury deposits at the north end of the Napa Valley. It mentioned "gabbro . . . along the contact between serpentine and volcanics." Moores got into his van, went to the Napa Valley, and looked. He then interested Steven Bezore, a graduate student, in working there. Bezore's master's thesis was the first demonstration of an ophiolitic complex in California, and led to the recognition of the Coast Range Ophiolite. After the winery, we stop at a crossroads store, Moores explaining that he requires coffee "to back-titrate the wine."

The descent is deep to the floor of the Napa Valley, which is flat. For a Coast Range valley, it is also spacious—as much as three miles wide. Vines cover it. Up the axis runs the two-lane St. Helena Highway, which seems to be lined with movie sets. This road is the vague but startling equivalent of the Route des Vins from Gevrey-Chambertin to Meursault through Beaune. The apparent stage sets are agricultural Disneylands: Beringer's Gothic half-timber Rhine House, Christian Brothers' Laotian Buddhist monastic château, Robert Mondavi's Spanish mission. Most offer tours, and wines to sip. As a day progresses, tongues thicken on the St.

Helena Highway, where the traffic begins to weave in the late morning and is a war zone by midafternoon. The safest sippers are in stretch limos, which seem to outnumber Chevrolets.

Most valleys in the Coast Ranges are smaller and higher than this one, their typical altitude at least a thousand feet. The southern end of the Napa Valley, being close to the San Francisco bays, is essentially at sea level. The valley floor rises with distance from the water, but not much. St. Helena, in the north-central part of the Napa Valley, has an elevation of two hundred and fifty-five feet. It is surrounded by mountains that are comparable in height to the Green Mountains of Vermont or the White Mountains of New Hampshire. Why this deep hole in such a setting?

The San Andreas family of faults is spread through the Coast Ranges, and outlying members are beneath the Great Valley. Where a transform fault develops a releasing bend—which is not uncommon—the bend will pull apart as the two sides move, opening a sort of parallelogram, which, among soft mountains, will soon be vastly deeper than an ordinary water-sculpted valley. In the Coast Ranges, most depressions are high and erosional. Some are deep tectonic valleys that are known in geology as pull-apart basins. In the Napa region, Sonoma Valley, Ukiah Valley, Willits Valley, and Round Valley are also pull-apart basins. Lake Berryessa lies in a pull-apart basin, and so does Clear Lake.

Where pull-apart basins develop—stretching and thinning the local crust, drawing the mantle closer to the surface—volcanic eruptions cannot be far behind.

In the Pliocene, after the Farallon Trench at this latitude ceased to operate and the San Andreas family appeared, the Napa basin had scarcely pulled itself apart
before the fresh red rhyolite lavas came up and air-fall
tuffs poured in. The Coast Ranges were aglow with sul-
phurous volcanism, its products hardening upon the
Franciscan. The nutritive soils derived from these rocks
prepared the geography of wine.

The rocks are known in geology as the Sonoma
Volcanics. Napa and Sonoma are Patwin Indian names:
"Napa" means house; "Sonoma" means nose or the
Land of Chief Nose. The rocks are the Land of Chief
Nose Volcanics. Chief Nose was a Tastevin before his
time. The heat of the volcanics lingers in the mud baths
and hot springs of Calistoga. The heat lingers under
cleared woods near Mt. St. Helena, where small power
stations dot the high ground like isolated geothermal
farms.

As the new fault system wrenched the country, fissures opened, and hot groundwater burst out in the
form of geysers and springs. They precipitated cryptocrystalline quartz and—in this matrix—various metals. Some gold. More silver. Near the surface, easiest to
mine, were brilliant red crystals of cinnabar (mercuric
sulphide). Mercury will effectively pluck up gold from
crushed ores. In the nineteenth century, the Coast
Ranges were tunnelled for mercury. It was carried
across the Great Central Valley and used in the Sierra.
The gold of the Coast Ranges was in those days in-
significant but is more than significant now. In the
nineteen-eighties, the Homestake Mining Company

dug two open pits in ridges north of the Napa Valley. In
surface area, they aggregate roughly a square mile. The
gold is too fine to be seen through a microscope but is
nonetheless there in sufficient concentration to be dissolved economically with cyanide. Discoveries of microscopic gold in California, Nevada, and elsewhere put
the United States in a position to surpass South Africa
in gold production by the turn of the twenty-first century—news that geologists regard as only slightly less
astounding than the landings on the moon. Homestake's underground mine in the Black Hills of South
Dakota is about a century old, and at latest count was
eight thousand and fifty feet deep—the deepest mine
in the Western Hemisphere. Homestake has produced
more gold than any corporation in North America.
With these new claims in the Coast Ranges, the company announced that it had more than doubled its re-
serves.

In 1880, Robert Louis Stevenson—aged thirty,
newly married, consumptive—fled the "poisonous fog"
of San Francisco and went into the mountains above
Calistoga, where he and his American bride and her
twelve-year-old son spent the summer squatting in an
empty cabin at a closed-down mine called Silverado.
From their high bench among rusting machinery and
rubbled tailings, they looked down into the green rec-
tangles of the Napa Valley.

The floor of the valley is extremely level to the roots of
the hills; only here and there a hillock, crowned with pines,
rises like the barrow of some chieftain famed in war.

Stevenson had more than a passing sense of the geology.

Here, indeed, all is new, nature as well as towns. The very hills of California have an unfinished look; the rains and streams have not yet carved them to their perfect shape.

Hot Springs and White Sulphur Springs are the names of two stations on the Napa Valley railroad; and Calistoga itself seems to repose on a mere film above a boiling, subterranean lake.

He began making notes for what became "The Silverado Squatters" and various settings for later work. He described the summit of Mt. St. Helena as "a cairn of quartz and cinnabar." He noted that Calistoga was a coined name. A Mormon promoter had been thinking of America's premier spa. Fortunately, his idea failed to travel, or there would be a Nevastoga, a Utastoga, a Wyostoga. Rattlesnakes resounded in the air like crickets. For a couple of months, Stevenson didn't know what he was hearing.

The rattle has a legendary credit; it is said to be awe-inspiring, and, once heard, to stamp itself forever in the memory. But the sound is not at all alarming; the hum of many insects, and the buzz of the wasp convince the ear of danger quite as readily. As a matter of fact, we lived for weeks in Silverado, coming and going, with rattles sprung on every side, and it never occurred to us to be afraid. I used to take sun-baths and do calisthenics in a certain pleasant nook among azalea and calcanthus, the rattles whizzing on every side like spinning-wheels, and the combined hiss or buzz rising louder and angrier at any sudden movement; but I was never in the least impressed, nor ever attacked. It was only towards the end of our stay that a man down at Calistoga, who was expatiating on the terrifying nature of the sound, gave me at last a very good imitation; and it burst on me at once that we dwelt in the very metropolis of deadly snakes, and that the rattle was simply the commonest noise in Silverado.

Without so much as a warning rattle, the owner of the Silverado Mine turned up one day, discovering and embarrassing the illegal squatter.

I somewhat quailed. I hastened to do him fealty, said I gathered he was the Squattee. . . .

Stevenson's summer was four years after the battle of the Little Bighorn. The West was that Old. Yet he counted fifty vineyards in the Napa Valley. Farmers had been in the valley for nearly half a century. In the eighteen-thirties, George Yount, of North Carolina, had been converted to Catholicism and had had himself baptized Jorge Concepcion Yount in order to obtain a Mexican land grant of almost twelve thousand acres. An English surgeon to whom the Mexicans also gave a Napa Valley land grant named his place Rancho Carne Humana. In 1876, the Beringer winery was founded by Germans from Mainz. In approximate replication of their ancestral home, they built Rhine House in 1883. The stretch limos park there now, beside wide lawns under tall elms. Off the jump seats come people who go inside and lay down forty dollars for the magnum opus *Beringer: A Napa Valley Legend.* Leafing through the book, Moores picks up the information that the foundation and first story of Rhine House are limestone. He goes outside and squints at the house through his ten-power Hastings Triplet. "Jesus Christ!" he says. For Moores, this is new ground. He has never before seen limestone that came out of a volcano. "It's poorly welded volcanic ash with lots of big vesicles, pumice lapilli," he goes on. "It's friable volcanic ash! A welded tuff! An ignimbrite!"

Louis Martini's cement-block roadhouse, south of St. Helena on the way to Rutherford, is a low, clean-lined, postroad-modern building that lacks windows and has a long portico and a few wrought-iron lamps. Its architectural statement is upper-middle prime rib. Among the building's tiled rooms are showcases of Martini wines and a long dark bar. No one hurries anyone away, and in the cool quiet we sample half a dozen bottles, talking geology with our noses in the outcrop. Louis Martini's wines are straightforward, stalwart, allusive, volcanic. They are prepared to travel—like the terrane they derive from, and like the first Martini (who emigrated from Italy in 1894), and, according to Moores, like Italy itself, which departed from Europe in the Jurassic but later went home. Italy became a prong of Africa, he says, cupping his hand and orbiting a cabernet sauvignon. Italy left Europe, joined Africa, and later smashed back into Europe in the collision that made the Alps. The quarried Tuscan serpentines in the walls of the Duomo and the Giotto campanile are particles of the ophiolites that underscore this story.

Martini's pinot noir has the brawny overtones of an upland Rioja, the resilient spring of an athletic Médoc. Moores wonders if I have noticed that "the claret coast of France" and the Cantabrian coast of northern Spain seem to suggest an open bivalve, with Bordeaux at the hinge. In the early Cretaceous, when the Atlantic was young and narrow, there was no water between western France and northern Spain; the hinge was closed. The whole of Iberia got caught up in the spreading, and was perhaps yanked by Africa as Africa moved northeast. A rift opened, and widened, and became the Bay of Biscay. In a comparatively short time, the Iberian Peninsula swung ninety degrees and assumed its present position.

During the zinfandel, Moores summarizes the United Kingdom as "the remnants of a collision that occurred at the end of the Silurian." Mélanges resembling the Franciscan were caught in it, he says—for example, Caernarvonshire and Anglesey, in Wales. Collisional ranges appeared, later to be dismembered by the opening of the ocean. In France, the Massif Central is actually a continuation of the northern Appalachians. The southern Appalachians go up to New Jersey and then jump to North Africa as the Atlas Mountains and then to the Iberian Plateau and to the Pyrenees, which were later enhanced by compressions that developed as Spain swung around.

During the Napa Valley Reserve Petite Sirah, I

mention the Brooks Range, where I have recently been.

The Brooks Range, Moores says, is a sliver of exotic continental material that came in from above Alaska, hit a subduction zone, and put ophiolite sequences along what is now the south slope. In the collision that followed, the exotic sliver was folded into mountains.

"When was that?"

"I forget. In the Jurassic, probably, or the early Cretaceous."

The Seward Peninsula—where Nome is, in west-central Alaska—is a piece of Jurassic blue schist surrounded by ophiolitic rock, but no one knows where the Seward Peninsula came from. For that matter, he adds, there is no certainty about where any of Alaska came from. It seems to consist entirely of exotic pieces that drifted to North American in Mesozoic time. South of the Denali Fault, which runs east-west and is close to Mt. McKinley, is the huge terrane that geologists call Wrangellia. It was an island arc, developed over an ocean plateau. Moores describes Mt. McKinley as "a bit of granite" that came up into Wrangellia after it arrived. Not long ago, Japan was attached to Asia. It drifted away. Japan is coming toward North America one centimetre a year. It may be a part of Alaska in eight hundred million years.

There is a shift change at Louis Martini's. One hostess replaces another. The new one says to her departing colleague, "Be careful out there. It's intense. They're driving all over the road."

That conversation in Louis Martini's winery occurred in 1978, when the theory of plate tectonics was ten years old and people who talked the way Moores was talking were widely considered daft. I may have thought it was the wine, but I was not in a position to know. Over the years, Moores and I have returned so often to the subject of world ophiolites and global tectonics—as they have recorded and described the changing face of the planet—that what follows is a sampling of all such dialogues, which I have compiled in the hope of reflecting, through his remarks, some of the geologic thinking of the nineteen-seventies, eighties, and nineties.

In order to move from place to place and let time float free, it would be well to bear in mind that the plate-tectonics narrative of the past fifteen hundred million years principally describes the assembling and disassembling of two supercontinents—Rodinia and

Pangaea. Of the mountain ranges of Rodinia we have nothing today but evidential roots, attended by some ophiolites that speak of the collisions that built those Precambrian mountains. After Rodinia breaks up, about six hundred million years ago, its fragments result in a map of the world so different from the present one that it could be a map of a different planet; Kazakhstan, for example, is contiguous with Norway and New England. By two hundred and fifty million years before the present, the scattered continents and microcontinents have reassembled as Pangaea, whose sutures are today expressed in dwindled but palpable topographic relief (the Urals, the Appalachians). While Pangaea in turn disassembles, in the Mesozoic, not only is the Atlantic Ocean born but all over the world recognizable pieces of dispersing land move in the direction of the present map.

Wherever tectonically emplaced ophiolites happen to be, they lead to local geographic histories within the general story of the successive supercontinents. The presence of an ophiolite is a notation that while something is added to a continental margin an ocean basin of unknown size disappears. It could be Pacific-size. Moores is planning a book relating ophiolites to their origins. Chapter 1 might develop his analogy between the great complexity of islands north of modern Australia and the loose landmasses that once cluttered the Farallon Ocean off western North America and are now consolidated as California and other additions to the continent. North of modern Australia is a confused piece of the globe, made so by the encroaching motions

of the Australian, Eurasian, and Pacific plates. They
have broken the crust between them into microplates
that look like the results of severe impact on a hard-
boiled egg. The small pieces continue to be rigid, and
remain in place, but the shell is shattered. The Philippine Plate, largest of the microplates, is surrounded by
ocean trenches six and seven miles deep. On the east,
Pacific crust is going under the Philippine Plate. On
the west, Asian crust is going into the Manila Trench,
where melting has produced the West Luzon arc and
where Taiwan is docking with mainland China. That
much is straightforward compared with the many
smaller microcontinents and minor ocean basins that
are also in the region, where a subduction zone is apparently in the process of flipping over, another is bending back upon itself, and another has curled around
almost far enough to meet itself in a circle. This is
a carnival of plate tectonics—of numerous island-arc-
to-island-arc collisions and continent-to-island-arc collisions. As in the story of western North America, some
arcs seem to be joining one another before attaching to
a continent.

Ocean crust of the Australian Plate descending
into the Java Trench has resulted in the arc from the
Andaman Islands to the Banda Sea: Sumatra, Java, Bali,
others. Where the Australian continental shelf has already jammed the trench and has picked up the Papuan
ophiolites, it has buckled its own Australian sediments
sixteen thousand feet into the air, making most of New
Guinea. Moores thinks that the subduction zone will
flip over now, and the Pacific Plate will begin to slide

under Australia. In that event, he says, "Australia will keep going and will pick up the Philippines and every intervening island and then go after Japan on Japan's way east."

"You're saying that a north-dipping subduction zone will swing like a pendulum and become a south-dipping subduction zone? That is possible?"

"That seems to be what's happening. That's what you see in the seismicity. There's nothing magical or indelible about the present plate margins. Consuming margins, especially, can change their nature very readily."

In the Sierra Nevada near the Mother Lode, where the geology suggests to plate theorists that a pair of ocean trenches came together in the Jurassic, evidently there was no spreading center, and the trenches just ate up the crust between them, leaving undigested the accretionary phyllites, cherts, argillites, and limestones that lie uphill from Auburn. To many people, the idea of unspreading seafloor being consumed from two sides by converging trenches has seemed especially farfetched. In the late nineteen-seventies, however, a pair of active trenches doing exactly that was discovered in the Celebes Sea. They are moving toward each other. The intervening crust is disappearing. With depth finders and seismographs, geologists can see this happening, but they can't explain it.

Tracing a finger northward on a geologic map of the world, Moores follows ophiolites from Beijing to Siberia. There are several parallel strings of them, connecting two Precambrian continental blocks. The place-names on his map are written in Cyrillic characters, because it is a Russian map, but the rocks are readable, in the international colors and symbols of the science. "These sutures tell you that China used to be separated from Siberia by two or three oceans," Moores says. "They disappeared in the Paleozoic."

North of China, the Verkhoyanski Mountains make a sinuous track through Siberia to the Arctic Laptev Sea. Landmasses extend two thousand miles east of the Verkhoyanskis and four thousand west, yet the mountains contain rocks derived from a spreading center in a vanished ocean. The Verkhoyanski ophiolites are lower Cretaceous in age—at least a hundred million years younger than the sutures in China, which were involved in Pangaea's assembling. The Verkhoyanski collision occurred after the supercontinent started to break up. As the Atlantic Ocean widened and the North American Plate moved west and the Eurasian Plate moved east, the two landmasses eventually touched each other, nearly halfway around the world, and made the Verkhoyanski Mountains. This was the plate boundary where Asia and North America actually came together. The Chukchi Sea and the Bering Sea, which separate Alaska and Siberia, are merely water lying on the North American continent.

Moores moves west to the Urals, which are flanked with ophiolites emplaced in Silurian time, in the middle Paleozoic, at the edge of the ocean that separated Asia from Europe. The ensuing collision did not begin until the Mississippian period, a hundred million years later. The fact that so much time passed between the emplacement of the ophiolites and the continent-to-continent collision means that a lot of seafloor was consumed, at least enough for an ocean a thousand miles wide. Russian geologists call this the Paleoasian Ocean. When the collision finally came, it completed Pangaea, two hundred and fifty million years ago.

Putting my hand on Spitsbergen and the rest of the Svalbard archipelago—ten degrees from the North Pole—I ask him, "What are they?"

With a sweeping move down the Atlantic he connects their story to Alabama. Among the various ocean basins that disappeared while Pangaea coalesced, the most intensively studied is the one that geologists have named for Iapetus, the father of Atlas: the ocean basin—or group of ocean basins—that lay between continental landmasses that are now substantial parts of Europe, Africa, and North America. Iapetus appears to have been larger than the modern Atlantic. Five hundred million, four hundred million, and three hundred million years ago, as Iapetus gradually closed, the lands on either side in no way resembled the modern configurations of Europe and North America, but they were composed of rock we see in those places now. In the Iapetus Ocean, or oceans, were arcs and trenches, spreading centers, microplates, subduction zones, strike-slip faults, a mess of islands. Much of this seems to have resembled the Farallon Ocean off California in Mesozoic time, and the southwest Pacific today. The collisions that eradicated Iapetus and made a kind of headcheese of the intervening islands began more or less at Spitsbergen, and—roughly, sporadically—crunched their way south. In the terms of the Old Ge-

John McPhee

ology, this was the series of mountain-making episodes that were known as the Caledonian, Taconic, Acadian, and Alleghenian orogenies. The trail of these events was blazed with ophiolites. Ophiolitic emplacements in Newfoundland, Quebec, and Vermont, for example, signal the docking of an island arc—the event long known as the Taconic Orogeny. The consequential mountain building in New England and much of eastern Canada was thought—in the early days of plate tectonics—to be the result of a continent-to-continent collision. "The Taconic Orogeny is a collision of ophiolitic terrane with the North American continent, full stop," Moores says. "It is an oceanic terrane—and not yet Europe—colliding with North America." In the assembling of New England, at least two more arcs followed. These Paleozoic additions to the eastern seaboard are remarkably analogous to the assembling of California an era later.

When continents collided, Africa docked with, among other places, the Old South. About a hundred and fifty million years later, when Africa departed, it apparently left a large piece of Florida, which is now covered with what Moores calls "a lot of modern limestones that developed on top of the Appalachian suture, which can be traced seismologically under northern Florida and off into the continental shelf."

Hesitating, I say to him, "Florida is covered with marine sand on top of limestone on top of Paleozoic rocks. The Paleozoic rocks derive from Africa. That is what you are saying?"

"That's right. Southern Florida is a piece of Africa which was left behind when the Atlantic opened up."

"People in Florida say that southern Florida is Northern and northern Florida is Southern."

"Civilization reflects geology."

Southeastern Staten Island is a piece of Europe glued to an ophiolite from the northwest Iapetus floor. Nova Scotia is European, and so is southeastern New-foundland. Boston is African. The north of Ireland is American. The northwest Highlands of Scotland are American. So is much of Norway.

While the Atlantic continues to open, the western Tethys, or Mediterranean Sea, is closing like a pair of tongs, hinged in the Atlantic crust roughly a thousand miles west of Casablanca. During the initial spreading of Tethys and later in the narrowing and even the lengthening of Tethys as effects of the forces that opened the Atlantic, the Mediterranean seafloor has been such a battleground that ophiolitic pieces of it are scattered around the basin like shell cases. Introducing chapters of the Mediterranean story, they are all through the Alps, Corsica, the Apennines, the Carpathians, the Dinarides, the Balkans, the Hellenides, Crete, the Cyclades, and the western part of Turkey.

The Mediterranean is full of tectonic rubble, no other single example being as large or as destructive as the Italian microcontinent, also known as the Adriatic Plate. Its western boundary is a subduction zone off the Campanian coast whose melt has become Vesuvius, and whose compressional distortions have become the

Apennines. The boundaries of the Italian microcontinent run north into Switzerland, northeast down the Rhine to Liechtenstein, east to include the Austrian Alps and Vienna, then south through Zagreb and Sarajevo and past the Vourinos Ophiolite in Macedonia to the central Peloponnesus, and back to the boot of Italy. In the Jurassic, the Italian microcontinent made its attempt to become a permanent part of Africa. As Africa moved northeast with the opening of the Atlantic, it returned Italy to sender, picking up the Tethyan ocean crust that became the ophiolites of the Alps. The Alpine collision began in the Eocene, about fifty million years ago, and has not completely stopped.

Muscat, in Oman, sits at the base of peridotite cliffs and is the only capital city in the world hewn into rock of the earth's mantle. Almost all of northern Oman is ophiolite, lifted by the shelf of Arabia in the closing of Tethys.

Four large parts of Africa, dating from the Archean Eon, are more than three thousand million years old: the West African Craton, the Congo Craton, the Zimbabwe Craton, and the Kalahari Craton. Long defined in geology as continental basements, continental shields, or continental cores, cratons are the ancient fundament to which younger and more legible rocks adhere. Plate tectonics now suggests (not to everybody) that the older parts of continents were themselves assembled much as the younger parts have been. For example, the African cratons are separated by belts of deformation that occurred after the Archean Eon but still in deep Precambrian time. Perhaps the deformed

rocks are suture belts where preexisting oceans disappeared. If plate tectonics was functioning then as it
functions now, the crusts of those vanished African
oceans consisted of rocks of the ophiolitic sequence.
There are late Precambrian ophiolites in the Kalahari
Desert of southwest Africa, in Sudan, in Egypt, in Arabia, and at Bou Azzer, in the western Sahara. Moores
says that Sudan, Egypt, and other parts of Africa are full
of exotic terranes—island arcs that were, as he puts it,
"crunched in there in the late Precambrian." He continues, "Geologists have long seen Africa as having developed in place, but the story must be wrong. Little is
known about the mobile tectonics there in Precambrian
time. I think it's kind of an unknown frontier of the
ophiolite story."

The mobile tectonics of more recent years are a
good deal easier to see. If you look on a world map at
Antarctica, South America, Africa, and Australia, you
virtually see them exploding away from one another.
You can reassemble Gondwana in your mind and then
watch it come apart. In the Cretaceous, Africa and
South America start to separate from Antarctica, and
from each other. India is still part of southern Africa,
but soon it breaks away. Australia remains attached to
Antarctica until the Eocene, when it breaks away, forms
its own plate, and heads north. India and Australia
move separately for a time but then weld themselves
together to become a single plate.

Madagascar begins to separate from Africa soon
after India does, but India leaves Madagascar far behind. The Seychelles move away from Africa in the

same manner—rifting obliquely, opening a small ocean basin with an active spreading center whose stairlike geometry (as in the Gulf of California) consists of short ridges connected by long transform faults. When the spreading stops, Madagascar and the Seychelles rejoin the African Plate.

I ask Moores why transform faults, like the San Andreas Fault, are so few and far between on land, whereas ocean floors are full of them.

He says, "They are rare on land because when they do appear in such a setting they rapidly take the land away and turn into marine transform faults. A transform fault carried India away from Africa. Look at the east coast of Madagascar. It's a long straight line, where India departed. Look at the corresponding part of India, the Malabar Coast below the Western Ghats. It's a long straight line. The Salinian Block, in California—with San Diego, Los Angeles, and so forth—will go on out to sea to the northwest, away from North America. The fault-divided halves of New Zealand's South Island, which were once apart, will come apart again. Equatorial Africa and northeast Brazil slid away from each other and developed an intervening sea."

I ask him why spreading centers and subduction zones take up most of the length of the world's plate boundaries and transform faults take up so little.

"Because the earth is a globe," he answers. "The curves of the spreading centers and the curves of the subduction zones will meet, or nearly meet. Where they don't meet, as in California, you find a transform fault."

When India separated from Africa, India's geographical center was more than a thousand miles farther south than the present position of the Cape of Good Hope. Three thousand miles of Tethys Ocean separated India from Tibet. India moved northeast as rapidly as any drifting continent in the calculable history of plate motions. At least one island arc lay in its path, and maybe several. There were microcontinents, too.

In and around the Himalaya are well-preserved ophiolitic sequences that describe the disappearance of that part of Tethys. These include ophiolites of Pakistan. Until the 1971 war, they were known in geology as the Hindubagh Complex. They are now called the Muslimbagh Complex. They run from the Indus Gorge east along the Indus River and the Tsangpo and Brahmaputra Rivers for two thousand kilometres, acquiring other local names along the way. This continuous belt of ophiolites consists of ocean crust that formed at a spreading center late in the Cretaceous and was emplaced on the northern margin of India in Paleocene time, when India had completed about half of its journey north. India evidently reached a trench with an island arc behind it, choked the trench, and picked up the ocean crust from the peripheries of the arc. That, in any case, is how Moores tells the story. He likens the ophiolite to a cow on a cowcatcher in front of an old western train. When those ophiolites were emplaced and India swept up the island arc (about sixty million years before the present), Australia was still on its own plate. After the two plates joined, in the

Eocene, the whole enterprise that geologists now refer to as the Australian Plate continued to move northward, gathering islands, for a few tens of millions of years. Then, with India as its hammerhead, it struck the Asian mainland. Moores thinks that the collision has scarcely begun.

This most emphatic of all contemporary continent-to-continent collisions is often described as a head-on crash, as if it had occurred within the ticks of human time. As India moved north, its highest rate of speed was a hundred and forty-two miles per million years. The present rate of compression is about a quarter of that, or two inches a year. If this could be recorded in stop-action photography, like the boiling swirls of cumulus clouds or the unfolding of a rose, it could indeed express itself kinetically. But two inches a year is an encounter so slow that a word like "collision" distorts its scale.

While India was closing with Tibet, it buckled the intervening shelf, raising from the sea a slab of rock more than a mile thick which consisted almost entirely of the disintegrated shells of marine creatures. From the depths of lithification to the rock's present loft, it has been driven upward at least fifty thousand feet. This one fact—as I noted some years ago—is a treatise in itself on the movements of the surface of the earth. If by some fiat I had to restrict all this writing to one sentence, this is still the one I would choose: The summit of Mt. Everest is marine limestone.

In the tectonic history of the globe, we have no idea how many times something of this proportion has happened. The probability is that it has happened often.

The boundary between the Eurasian Plate and the plate that carries India and Australia seems pretty obvious, but actually it cannot be narrowly defined. It is not as simple and precise as the Indus suture, where ophiolites are embedded along the northern slopes of the higher mountains, and it cannot be limited to the Great Himalaya Range itself, though that appears to be a clear partition between the hitter and the hit. Across two thousand miles, from the Ganges River north to Lake Baikal, the boundary between the Australian Plate and the Eurasian Plate is indistinct. It was once described as a separate plate, and was unfortunately named the China Plate. The whole zone is seismically active. It contains the highest large plateau in the world. The Indian collision has produced additional mountain ranges north of the Himalaya that are comparable in altitude to the Andes. Included also is the Sinkiang Depression, where collisional downbending has put the ground below sea level. Essentially, all of China is a part of the plate boundary, and in all of China is a part of the plate boundary, and in all of China there are very few rocks that are undeformed. Chinese geologists travelling in America incessantly snap pictures of simple flat-lying sediments—a geological basic that they have seldom seen. The crust under the Tibetan Plateau is twice as thick as most continental crust. The Australian Plate, pressing northward, seems to have caused this. To accommodate two inches of relentless annual advance, various things have to bulge or give way. The mountains have risen. The plateau has thickened. But these two changes have not been enough to account for the total compression. A growing number of geologists, following work that is being done by a group of French tectonicists, are beginning to agree that a large part of Southeast Asia has also been forced to one side. Where Burma meets India, the high ranges bend almost at a right angle and go off to the southeast. This is the Burma Syntaxis (the term refers to a bend in a mountain chain), and near it are the beginnings of a whole series of great rivers—the Brahmaputra, the Mekong, the Irrawaddy, the Salween—initially in parallel valleys, veining Southeast Asia from the Bay of Bengal to the South China Sea. Controlling these valleys are long strike-slip faults, in motion like the San Andreas. The French tectonicists are proposing that Vietnam, Laos, Thailand, Cambodia, Burma—the whole of Indochina—slid southeastward among those strike-slip faults, like a great terrestrial hernia. On a relief map you can see India ramming Asia and squeezing all that country out to the southeast. As the mechanism has gained acceptance among tectonic theorists, it has become known as continental escape.

I ask Moores if he thinks there's a chance that plate tectonics may someday seem to have been a rational fiction, as the geosynclinal cycle does now.

"For parts of the world, maybe so," he says. "Whatever is going on in central Asia is no one's idea of plate tectonics. But as an explanation for eighty per cent of the surface processes of the earth, plate tectonics is in, firm." Repeating the words of the volcanologist Alex

McBirney, he says, "Remember, 'In the next ten years, our confusion will reach new heights of sophistication.'"

(Or, in words dubiously attributed to Mark Twain: "Researchers have already cast much darkness on the subject, and if they continue their investigations we shall soon know nothing at all about it.")

The thought occurs to me, not for the first time, that I am following a science as it lurches forward from error to discovery and back to error. In my effort to describe some of the early discoveries of plate tectonics, I must also be preserving some of the early misconstructions.

"Inevitably," Moores agrees. "That is the nature of science, and geology is surely no exception." His mentor Harry Hess, a combat veteran with the rank of rear admiral in the Naval Reserve, once told him, "Geologists make better intelligence officers than physicists or chemists, because they are used to making decisions on faulty data."

The data that sketch departed geographies are actually numerous. Where pictures are clearest, the data cross-check with confirming frequency. For example, the ophiolitic narrative will conform with ancient latitudes preserved in the remanent magnetism of rock. The fossil record must not disagree. Where strike-slip faults have sliced a landscape and carried two sides apart, matchups can be traced in time and space. Sedimentary sequences, blue-schist belts, batholithic belts, thrust belts, and mélanges will orchestrally tell what happened. If they are not synchronous, it didn't happen.

That Asian plate boundary two thousand miles wide untidies the theory of plate tectonics more than any other place in the world with the probable exception of the American West. Moores is among the growing number of geologists who believe that Salt Lake City is on the eastern side of a muddled and dishevelled boundary between the North American and Pacific plates. Not long ago, when the boundary was utterly different—when an ocean trench off North America was consuming the Farallon Plate—the Rocky Mountains appeared, from Alaska to Mexico. In a major way, they defy explanation. As you look at the world map and see India hitting Asia, with the Himalaya and all the additional deformation to the north to show for it, you might begin to wonder why there is no India against the west coast of the United States. Obviously, the Smartville Block and the other accreted arcs added something to the mountains' compression, but the impact of those terranes could not have been sufficient to deform a third of a continent. Force from the west evidently made the mountains. But what force? If it was a colliding landmass, where has it gone?

If you run your eye up the coast a couple of thousand miles, you see protruding from North America a body of land at least as conspicuous as India. Staring at the map one day, I ask Moores if he is ready for some academic arm waving on a windmill scale.

"Shoot," he says.

"Why isn't Alaska the missing India of North America?"

"It is."

"Why didn't Alaska—not all at once but in parts and in successive collisions—strike North America at the California latitudes and then take off on transform faults and slide north to where it is now?"

"That's exactly what Alaska has done. When the Brooks Range swung into place, none of the rest of Alaska was there. Below the Brooks Range, all of Alaska consists of exotic terranes. They seem to have come from the Southern Hemisphere, and even from the western Pacific. Where each and every part originated we have no idea, but a lot of it collided down here in California and then went north on transform faults. Sonomia, Smartville, and so forth are probably fragments of terranes that in large part broke away and went north."

From radiometric dating, paleomagnetism, matching fossils, and connectable orogenies, George Gehrels, of the University of Arizona, and Jason Saleeby, of the California Institute of Technology, have proposed that the Alexander Terrane of southeastern Alaska, which includes Juneau and Sitka, drifted ten thousand miles from eastern Australia to Peru and then north to its present position. Vancouver Island seems to have followed; its paleomagnetism indicates that it came from the latitude of Bolivia and arrived in the Eocene.

Such travels seem modest in comparison with a vision that Moores has of the world half a billion years earlier. In 1991, he published a paper claiming that Antarctica and western North America were once conjunct. This was during the existence of Rodinia. Tectonicists have for some time agreed that something must have rifted away from western North America in latest Precambrian time—that the craton cracked and separated, much as the Nubian-Arabian Craton is rifting now and making the Red Sea. In proposing Antarctica as the other side of the rift, Moores traces Precambrian lithologies of eastern Canada down through Alabama, Texas, and Arizona into Queen Maud Land, East Antarctica. Ian Dalziel, of the University of Texas, who had made a field trip with Moores to Antarctica, took Moores' proposal and extended its Precambrian juxtapositions, reconstructing the whole of Rodinia. According to Dalziel and Moores, if you had journeyed due north from Morocco you would have crossed western Africa and gone into Venezuela and through Brazil and Chile to West Virginia and on through Arizona into Antarctica, beyond which lay Australia. When they published their conclusions, *Time*, *Science News*, the *New York Times*, the *Los Angeles Times*, the *San Francisco Chronicle*, the *Washington Post*, and countless other publications adorned the story with maps beside which the cartographic efforts of the fifteenth century seem to be precision documents.

Running a finger down through Mexico and into Guatemala almost to Honduras, Moores says that an east-west fault zone there is laced with ophiolitic rock and seems to have been the southern extreme of Paleozoic North America. After Rodinia dispersed, what lay beyond Guatemala was open ocean. When Pangaea later coalesced, north and south touched there.

One of the places where the breaking up of Pangaea may have begun, in the Mesozoic, was not far

away. If you reassemble the Triassic terranes around
the Atlantic, the dikes radiate from a point in the Ba-
hamas—suggesting that a hot spot that was centered
there broke open a large part of Pangaea. As the supercontinent rifted, a large oceanic gap reopened between
North and South America—in effect, a western exten-
sion of Tethys. It contained no Antilles, Lesser or
Greater—none of the present island arcs or subduction
zones of the Caribbean. It was blue, abyssal ocean. Its
bottom was ocean crust-and-mantle—the ophiolitic sequence. Evidently, the Caribbean Plate, bringing arc
sequences with it, came drifting in from far to the west
to take up the position it holds today. It appears to have
collided with the Bahama platform, and perhaps with
the shelf of North America, in latest Cretaceous time.
Ocean crust chipped off its edges as it fitted into place.
All around the basin, island arcs came up above sea
level, and fragments of ocean crust came up with them,
as the ophiolites of northern Venezuela, of Cuba, of
Hispaniola, of western Puerto Rico. The island of
Margarita, near the Venezuelan coast, appears to be
one large ophiolite. Off the Cayman Trench, in mid-
Caribbean, the ocean crust is thicker than ocean crust
commonly is. Thick basalts seem to have poured out
over the original ophiolitic sequence in some sort of
mid-plate volcanic event. "The only other place we
know about where that sort of thing occurred in that
time is out in the western Pacific, in the Nauru Basin,"
Moores continues. "Some people have suggested that
these two things may be related. If you do a reconstruc-
tion of the East Pacific Rise and the plates in the Pa-

cific, you can bring the Caribbean back into contact with the Nauru Basin in early Cretaceous time." If that is where the Caribbean Plate came from, the distance that it travelled eastward is eight thousand miles.

The gap between the Americas was subsequently filled in two stages. Not long after the arrival of the Caribbean Plate, Honduras and Nicaragua drifted in from who knows where. Ultimately, Panama came in from the Pacific, about seven million years ago. The collision calls to mind a cork going into a bottle, but was not so neat a fit. Panama was a part of an island arc, and the ophiolite that preceded it runs from Costa Rica to the Colombian Cordillera Occidental and on down to the Gulf of Guayaquil, indicating that the Choco Terrane, as the arc is called, was nearly a thousand miles long.

In Colombia's Cordillera Central, one range east of the Choco Terrane, are older ophiolites that report like a tree ring the continent's growth in successive laminations. Moores has been in Brazil among Precambrian ophiolites that seem to describe the basement of the continent collecting. In southern Patagonia, he has traced a line of ophiolites, called the Rocas Verdes Complex, from the fiftieth parallel to Tierra del Fuego. The Atlantic island of South Georgia, two thousand miles east of Tierra del Fuego, is part ophiolite and is believed to be a travelled piece of the Rocas Verdes Complex.

"Travelled?"

"Probably on a transform fault," Moores says.

In other words, South Georgia was once the easternmost part of Tierra del Fuego, and it took off.

Another part of Tierra del Fuego appears to have departed but returned. That, at any rate, is what Moores and others concluded in 1989 during a geologic voyage there. The Straits of Magellan lie across an ophiolite complex that they interpret as the back-arc basin behind a piece of South America that moved west before changing its mind. Perhaps South Georgia will come back, too.

Down the high Andes from Ecuador to fifty degrees south are no known ophiolites. This is a surprising interruption in a story that touches every other plate-boundary mountain range or former-plate boundary mountain range in the world. After being tucked into the tectonics all the way from Alaska to Ecuador, ophiolites disappear. The gap in which they are missing is equal to the distance from the equator to Seattle. In ophiolite tectonics, perhaps the largest question at present is: What has happened in the central Andes?

In his office in Davis, with the Americas spread out on paper before him, Moores greets the question laconically. He says, "It makes you wonder."

I remark, "It's the one place on the western margin of the Americas where you don't find ophiolites providing some sort of story about exotic terranes—about the Smartville Block joining California, about Caribbean crust coming several thousand miles to its present position. Your intuition must be telling you something."

Moores says, "There are two possibilities: one, it is different from everywhere else that we know of; two, the evidence is hidden or is not preserved. My guess— this will make Andean geologists cringe—is that much of the Mesozoic volcanics one sees in the western Andes and the coastal ranges of Chile and Peru represents something that was at one time exotic to South America. My guess is that there's a suture. We just haven't found it. It may well be on the Chilean-Argentine frontier."

"So you're talking about an ophiolite announcing a terrane that came drifting in from wherever to fill the four thousand miles from the equator to fifty south, and you don't even know where the ophiolite is?"

"That's right."

"So you're . . ."

"Doing violence to the geology as it's known."

"The understanding of plate-tectonic history as it appears to be developing for the Pacific margin of the two Americas from Alaska to Tierra del Fuego includes lots of evidence for such terranes attaching themselves all the way, with the single exception of . . ."

"Zero to fifty south."

"And you think that . . ."

"Chile is exotic terrane."

"Chile and western Peru are from somewhere off in the South Pacific?"

"It's not a credible position to defend. But that's what I think."

"Where are the ophiolites on the east side of the Andes?"

"They haven't been found."

We go out of Davis one morning past a sign that says "OPEN TRENCH," and head for San Francisco on Interstate 80. In the median are dense effusions of oleander, in blossom pink and white, beguiling the westbound traffic with the pastel promise of California. A pickup in front of us is carrying an all-terrain vehicle studded with flashing sequins. As we move out to pass, the pickup abruptly moves out and blocks us. The pickup has California plates and a bumper sticker. "Don't Like My Driving? Dial 1-800-EAT-SHIT."

By the edge of the western hills, Mt. Diablo stands up like a hat on a table. All the way from Sacramento to the Coast Ranges you can see it from the highway. Moores calls it a piercement structure. It is a balloon-like mass of Franciscan mélange that has been squeezed up through valley sediments as if from a pastry sleeve. It is nearly four thousand feet high. Its top-ographical base is at sea level, beside the common delta of the great conjoining rivers, where a canal with no locks takes oceanic merchant ships across the Great Central Valley to Sacramento.

To climb Mt. Diablo, you go up through Pittsburg out of Honker Bay. In the tranquillity of its oak woodland (the mountain is a park) you see the rhythmically bedded red cherts characteristic of the Franciscan. You see its featureless, unbedded sandstones. You see elements of the Coast Range Ophiolite. Moores calls the mountain "low-cost fruitcake, stirred up even more than most Franciscan." Interlayered volcanics and cherts are folded and refolded there like the leaves of a croissant. From the summit, you can see more than a hundred miles to the High Sierra or look down into a quarry cut in sheeted diabase from some ocean's spreading center who knows where.

The geology along the interstate west-southwest of Davis recapitulates the traverse to the Napa Valley, but in less obvious fashion, for the hills are lower; the delta and the bays are in a structural depression. In the swells of the oak woodland, the bent-upward sediments of the Great Valley are under the straw-brown grass. Where the grass is thin, you can see the bedding. Near Cordelia, Green Valley Creek comes in from the north, following the Green Valley Fault, which is actively slipping. Here in the hill country between Fairfield and Vallejo, the lovely hummocky topography is the result of creeping landslides, earthflow, solifluction. The solifluction scars are like stretch marks, waves advancing downhill. This is not the sort of place to site a house, but it is the sort of place where houses are sited. Here geology in motion is just another factor in daily life. When a classified ad in a local paper says "Owner Suddenly Called East Must Sell," the possibility is not inconceivable that the house is a sled.

On the top of Sulphur Springs Mountain, before Vallejo, the highway is in a benched throughcut a hundred feet high. On 1-80, this is the beginning of the Franciscan. The Coast Range Ophiolite is here as well. Some of the rock is serpentine. We pull over, and walk the cut from one item to another in the mélange. After the serpentine comes a black wall that reflects light as if it were made of obsidian. Moores opens a pocketknife and easily carves the rock. He studies it in his hand lens. He says it is "the decrepitated matrix of the Franciscan"—the scaly clay in which are embedded all the continental shards and abyssal sediments, the bits of seamounts and ocean crust, the litter of half the world.

The interstate descends toward Vallejo, toward Benicia, toward Suisun and San Pablo bays. When Robert Louis Stevenson first saw Vallejo, he described the community as "a blunder." Vallejo was twice, briefly, the capital of California. Benicia was Mrs. Vallejo. Between the two bays, we cross Carquinez Strait and immediately pass through the soft marine sediments of the largest roadcut on Interstate 80 between the Atlantic and Pacific Oceans. Its vertical dimension is three hundred and six feet, but the sediments is so weakly cemented that the two sides lie open like butterfly wings and are thus immense. After the Farallon Trench quit, these marine deposits settled upon the Franciscan mélange. Throughout the Coast Ranges, the mélange has an icing of sediment, acquired while it was still underwater.

Running through Richmond over landfilled marshes beside San Pablo Bay, we cross the Wildcat Fault and, moments later, the close and parallel hayward Fault. Off to our left, southeast, runs the long escarpment of the Berkeley Hills. Their steepness breaks at the Hayward Fault, below which is gently sloping ground. The obeliscal Berkeley Campanile, of the University of California, stands like a marker close by the fault, whose itinerary on the campus goes right through Memorial Stadium. The Hayward Fault is like one of the yard lines.

The road veers west, and we are suddenly high above water, on the upper deck of the San Francisco–Oakland Bay Bridge. Left and right of us are fifty miles of safe anchorage, waters that are in places twelve miles across. It strains credulity that in two centuries of nautical exploration this most prodigious harbor in North America did not reveal itself to a single ship. So far as is known, no ship passed through the Golden Gate until 1775—a modern date, even in California time.

As a geologic feature, the bay is the youngest thing in sight—younger than the rivers that feed it. During the glacial pulses of the Pleistocene, when so much water rested on the continents as ice, sea level was lower than it is now by a couple of hundred feet, and shorelines worldwide were far outboard of present beaches. Past New York City, the Hudson River kept on going nearly a hundred miles east before it reached the shore

of the Atlantic. Similarly, the Sacramento–San Joaquin
went out through the Golden Gate and forty or fifty
miles west before it emptied into the sea. When the ice
melted, the sea came up, and drowned innumerable
river valleys—drowned the Susquehanna and made
the Chesapeake, drowned the Delaware to Trenton,
drowned the Hudson to Albany, drowned the Sacramento–San Joaquin from the Golden Gate through the
Coast Ranges and into the Great Central Valley, filling
the Bay Area's bays.

If Alcatraz, Angel Island, Yerba Buena, and so
forth were elsewhere in the Coast Ranges, they would
be the summits of mountains, and not islands in a bay.
Something has depressed the bay region, incidentally
allowing the rivers to get out of the Great Valley. In
breadth and depth, the depression is unlike any other
among the mountains of the California coast. Asked to
explain, Moores can only speculate. Possibly the depression is just a system of erosional valleys, at the business end of the state's great watershed. Maybe it's a
synclinal fold—a compressional trough, made as a side
effect of San Andreas motion. Maybe it's a huge pull-
apart basin. Maybe it's all three. Of his guesses, he likes
best the pull-apart basin.

"What's under the bay?"

"Volcanics, conglomerates, glaucophane schist,
sandstone, serpentine, chert—Franciscan. This is the
place it was named for, and this is the tectonic product.
The Franciscan is under the city of San Francisco and is
the basement of the bay. Franciscan rocks in the Bay
Area are from more parts of the subduction complex

than you are likely to find in such concentration anywhere else on the coast."

The bridge arrives at Yerba Buena Island, and a tunnel runs through it: through sandstone of the Franciscan, outcropping above the entrance and derived from some continent somewhere, perhaps from our own—indurated American sediment that slid down the slope to the trench and there became incorporated into the Franciscan. In the light again and on the western span of the bridge, the run is short to the white city.

Not just any city is the topic of a serious tourist guidebook called *A Streetcar to Subduction,* a copy of which Moores and I have with us. Written by the geomorphologist Clyde Wahrhaftig, of Berkeley, it is a shining little book of geological field trips on public transportation.

Not just any city can claim to have formed in a trench where the slab of a great ocean dived toward the center of the earth, where large pieces of varicolored country came together, and where competent rock was crushed to scaly clay. After the churning stopped and the whole mixture was lifted into the weather, the more solid chunks very soon stood high and the softer stuff washed down. In Ina Coolbrith Park, we climb to the top of Russian Hill. Lining the path are large sharp pieces of red Franciscan chert, but Russian Hill is sandstone and the chert is decorative. It has come in trucks. Moores trains his binoculars on Alcatraz. He has read the geology of Alcatraz. He says it is a sequence of quartzofeldspathic sandstone and shale. In the compote of the Franciscan, Alcatraz is lying on its side. From

Russian Hill he can see that—in the bedding planes. The rock of Alcatraz is continental in origin. There is no telling what continent, or when it escaped. The hill we stand on is also quartzofeldspathic sandstone and is believed to be of the same provenance as Alcatraz. Behind us is the summit of Nob Hill. Russian Hill and Nob Hill are actually the mammary climax of the same small mountain. Some tectonicists look upon this sandstone as being so discretely integral that it deserves nomenclatural distinction. They see it as a picocontinent that drifted into the Franciscan. In their terminology, Nob Hill is part and parcel of the Alcatraz Terrane.

Because the mélange contains rocks from all over the Farallon rim, it lends itself to reclassification by those who prefer to describe it not as a tectonic unit but as a boneyard of exotica. To them the Bay Area is not so much the type locality of the Franciscan as a tessellation of six miniterranes. Over the water past Alcatraz we see Angel Island and the southern tip of the Tiburon Peninsula, which, with scattered bits along the Hayward Fault in Richmond, Berkeley, and Oakland, have been called the Yolla Bolly Terrane. The Marin Headlands Terrane includes not only southern Marin County but much of the city of San Francisco. There is a San Bruno Mountain Terrane, a Permanente Terrane, and, where doubt retains a foothold, an Unnamed Terrane.

We cross the Golden Gate, take the first exit, and curl downhill in overshadowing roadcuts of radiolarian chert. Radiolaria are creatures that live near the tops of warm oceans and look like microscopic sea urchins. After they die, their external skeletons go to the bottom to be radiolarian ooze, which lithifies as a very hard, very beautiful rock—a wine-red cryptocrystalline quartz of arrowhead quality. In fifty per cent of the world ocean, radiolarian chert lies like an enamel on the ophiolitic sequence. Where the sequence goes, so goes the chert.

Where you find chert from the open ocean, if you keep going downsection, basalt will soon follow, Moores remarks. Walking downsection, he finds the contact of red upon black. He says that the basalt was a seamount, one of the countless submarine volcanoes that ride the ocean floor.

In the early nineteen-thirties, the north pier of the Golden Gate Bridge was sunk two hundred feet through the red chert and into the basalt below it. At the north pier, the competence of the rock was never a problem. Support for the south pier, in San Francisco, was a good deal less promising. The south pier had to stand, in water, on sinuous, slippery serpentine, the state rock. Also under the ocean, two miles away, was the San Andreas Fault. The serpentine was thought to be potentially unstable, so it was hollowed out, like a rotten molar. The hollow was a little more than an acre, and ten stories deep. It was to be filled with concrete to anchor the bridge. While it still lay open and dry, within coffering walls thirty feet thick, the structural geologist Andrew Lawson, of Berkeley, was lowered in a bucket to inspect the surface of the bedrock. With his pure-white hair, his large frame, his tetragrammatonic mustache, Lawson personified Higher Authority. The stability of the serpentine had been called into question and made a public issue not only by a mining engineer but by the world-renowned structural geologist Bailey Willis, of Stanford, who predicted disaster. Lawson regarded Willis' assessment as "pure buncomb." Getting out of his bucket a hundred and seven feet below the strait, Lawson found that "the rock of the entire area is compact, strong serpentine remarkably free from seams of any kind." He wrote in his report, "When struck with a hammer, it rings like steel."

About the proximity of the great fault, Lawson had realistically observed (during the design phase) that an earthquake strong enough to knock down the bridge would also raze the city. He went on to say, "Though it faces possible destruction, San Francisco does not stop growing and that growth necessarily involves the erection of large and expensive structures."

In June, 1935, when the south tower stood nearly complete—seven hundred and forty-seven feet high, with no cables attached—it began to sway in a middle-energy earthquake. A construction worker named Frenchy Gales—as quoted in John van der Zee's *The Gate*—continues the story:

It was so limber the tower swayed sixteen feet each way. . . . There were twelve or thirteen guys on top, with no way to get down. The elevator wouldn't run. The whole thing would sway toward the ocean, guys would say, "Here we go!" Then it would sway back, toward the Bay. Guys were laying on the deck, throwing up and everything. I figured if we go in, the iron would hit the water.

The iron did not hit the water. Charles Ellis, the chief designer, had likened his developing bridge to a hammock strung between redwoods. Addressing the National Academy of Sciences in 1929, he said, "If I knew that there was to be an earthquake in San Francisco and . . . this bridge were built at that time, I would hie me to the center of it, and while watching the sun sink into China across the Pacific I would feel content with the thought that in case of an earthquake I had chosen the safest spot in which to be."

Moores and I recross the bridge. South of the south pier, we go down a steep trail to the Pacific beach that comes up from Point Lobos. There are serpentine outcrops above the beach—large blocky hunks in the scaly Franciscan matrix. The closer we come to the pier of the bridge, the more serpentine we see. "It could be part of the basement of the Marin Headlands Terrane," Moores says. "The seamount, with this serpentine as a basement, is a really far-travelled piece. It's old and equatorial. It began life way out in the central ocean. What a pile of hash-mash to put a bridge on!" The serpentine is massive, soft and soapy, threaded with asbestos, and below the great bridge it stands up in cliffs. In high positions are concrete gun emplacements built for five-inch pillar-mounted rifles, for six-inch disappearing rifles. There are nudes on the beach. Other men on the beach are sitting upright in little pillboxes that shelter them from the wind. The traffic on the bridge hammers the expansion joints and sounds like the firing of distant guns.

In section more than a mile wide, the serpentine traverses San Francisco from the Golden Gate Bridge to the old naval shipyard on the bay. It underlies all or parts of the Presidio, the University of San Francisco, the Civic Center, Mission Dolores, Haight-Ashbury, Pacific Heights, Hayes Valley, Bayshore. Potrero Hill is serpentine. On Candlestick Hill, near the naval shipyard, we find pillow lavas and red chert. Climbing Billy Goat Hill, above Thirtieth and Castro, we see more pillows under more red chert. Behind a grammar school in Visitacion Valley, at the south end of the city, we scramble up a dark hill of gabbro. Around a pond in McLaren Park, nearby, are diabase boulders.

Chert, pillow lavas, diabase, gabbro, serpentine— item for item, the city seems to be in part composed of the ophiolitic sequence. But here in the Franciscan it is not a sequence. "These rocks are not necessarily from one ophiolite," Moores comments. "They are probably from all over the globe—bits of ocean crust, of varying age and provenance, collected in the mélange."

It was Andrew Lawson who, in 1895, named the rocks Franciscan. He assumed that they were a conventional formation with traceable stratigraphy—with an eroded structure that could nonetheless be deciphered and spatially reconstructed. One might as well empty a cement mixer and try to number the pebbles in the order in which they entered the machine. On the other hand, Lawson came uncannily close to describing the ophiolitic sequence and to seeing it as ocean crust, and in this respect he was more than half a century ahead of his time. Of the San Francisco sea cliffs that are now described as pillow basalts, Lawson wrote that they resemble "an irregular pile of filled sacks, each having its rotundity deformed by contact with its neighbor." In

1914, enumerating the igneous rocks of the Franciscan, he came closest to describing the ocean-crustal sequence: "The igneous rocks are genetically allied peridotites, pyroxenites, and gabbros, the first named preponderating and being generally very thoroughly serpentinized; and the second spheroidal and variolitic basalts and diabases." He mentioned sandstone that had been deposited "upon the sinking bottom of a transgressing sea," and added that "cherts followed." Lawson's sinking bottom was the cooling ocean slab, on its way to the trench.

There are hillsides in downtown San Francisco that are too steep for cars, cable cars, or human locomotion short of rope-and-piton climbing. If you look at a street map, you will see a hiatus in Green Street near the Embarcadero. A few feet west of Sansome, Green ends at the edge of a vacant lot. No one can build there, because the vacant lot is vertical, a cliff of solid rock. "Grungy-looking sandstone," Moores remarks, glancing upward. "Can't see any bedding. It's marine sandstone that went into the trench." Houses line the top of the cliff. An apartment building with cantilevered balconies seems to hang over the edge. Somewhere up there, Green Street continues west.

Filbert Street, a couple of blocks away, is similarly interrupted. The escarpment is a little less sheer. A roadway is out of the question, but the street turns into a staircase. Houses are on both sides of the steps, some of them dating from the mid-nineteenth century. A few are hutlike, with stovepipe chimneys and sagging windows. It is reported that in April, 1906, when the temblor destroyed the municipal water system and the helpless city was being razed by fire, these houses were saved by sheets, blankets, tablecloths, and bedspreads soaking wet with wine. The pale painters and coughing poets who once lived here are gone now, replaced by readers of *Barron's Weekly*. Their cliff dwellings are charming past the threshold of envy, and they look ten miles over water.

We climb three hundred and eighty-seven risers, not to mention intercalated terraces and ramps. We are climbing, actually, Telegraph Hill. Moores refers to it as "a single large thick turbidite bed." In the rock behind the stairway, marine sandstones are interbedded with shales. The sands and muds probably derived from North America but slid out far enough to get into the trench.

At the summit of the hill is Coit Tower, and we go on up that as well—for a dollar, in an elevator. Coit Tower is two hundred and ten feet tall, and its observation terrace is five hundred feet above the sea. The view incorporates the city, the bays, the shoreline suburbs. We look down upon absurdly straight thoroughfares roller-coasting the precipitous hills. Moores says, "The hills have risen rapidly and have therefore eroded steeply. They're still rising rapidly. San Francisco streets were drawn on paper, without regard to geology or topography. There is one reaction. You laugh."

Skyscrapers ascend the apron of Nob Hill. South, from the tower, the view passes over them as if they were stalagmites. The eye is stopped by the San Bruno Mountain ridgeline, altitude thirteen hundred feet, fencing off the peninsular city. San Bruno Mountain, like Telegraph and Nob and Russian hills, is a large loose piece of marine sandstone within the Franciscan mélange. In the opposite direction, the view crosses the Golden Gate, passes over the Marin Headlands, and is again stopped, this time by Mt. Tamalpais, another block of float sandstone, twice as high as the San Bruno ridge. Francis Drake, the English pirate, two years shy of being knighted by the queen, spent the winter of 1579 camped beside a Pacific beach close to the base of Mt. Tamalpais. Neither he nor anyone in his crew went up the mountain to look around, to discover the three hundred and fifty square miles of protected water close to their encampment. The probable explanation is fog: the cold and almost quotidian sea fog that will overlap the coastal land when the air of California is otherwise cloudless; the fog that fosters the growth and survival of redwoods; the fog that conceals the Golden Gate Bridge and brings out the sounds of tubas.

In the summer and fall of 1769, sixty-four Spanish soldiers walked north four hundred miles from San Diego in search of Monterey Bay. Having no more to go on than a navigator's description a hundred and sixty-six years old, they failed to recognize Monterey Bay; and they kept on walking, another hundred miles, until they came up against a large coastal mountain. On a clear day, they climbed it. They were fourteen miles from the Golden Gate, but they could not see water there. All they could see was the ocean to their left and, ahead of them, an endless reach of mountains. Forty

miles up the coast they could see Point Reyes. Several
soldiers were sent ahead to blaze a trail to Point Reyes
and prove that it was not Monterey. Their exact route is
not clear. It is more than probable that they went far
enough to be stopped by the deathly currents of a narrow strait set against unfriendly cliffs, and that—from
one of the numerous hills—they became the first Euro-
peans to look upon San Francisco Bay. Soldiers back at
the main encampment, nearly starving, climbed San
Bruno Mountain hunting deer, and saw the bay.

We, on the tower, have the city of San Francisco
spread around us, whereas the soldiers, from similar
positions, looked out upon a confused topography of
deep swales, creased gullies, and high dunce-cap hills.
The hills were all but treeless. They were matted with
bush monkeyflowers, woolly painted cups, coffeeber-
ries, Christmasberries, bush lupine, and poison oak. In
the swales and gullies were wax myrtle, arroyo willows,
coast live oaks, creek dogwood. Large parts of the future city were covered with marching dunes, restrained
by dune tansy, coyote shrubs, and sand grass. The
scouts probably shrugged. In any case, their mission
had failed.

Seven years after the discovery of the bay, thirty-
three families with the intent of settling beside it
trekked eight hundred miles north from the part of
Mexico that now is Arizona. Their leader, Juan Bautista
de Anza, went ahead of them and examined the terrain.
On top of the serpentine cliffs quite close to what is
now the southern approach of the Golden Gate Bridge,

he erected a cross. This was to be the site of the citadel (*presidio*). He erected another cross a few miles away, beside a small lake. This was to be the site of the mission. When the settlers arrived, they pitched their tents by the lake. Their seventh day of residence was the Fourth of July, 1776.

S an Francisco is on the North American side
of the San Andreas Fault, barely. The fault comes in
from the ocean at Mussel Rock and goes straight down
the San Francisco Peninsula, southeast. It runs close
beside Skyline Boulevard above South San Francisco,
San Bruno, Millbrae, Burlingame. The "skyline" is one
of several Coast Range ridges that are separated from
others by the depression of the bay. Five miles south of
the city proper, Moores and I left Skyline Boulevard
one December day and climbed through a subdivision
to an elevation that was close to a thousand feet. We
walked a hundred yards or so to a lookoff above a finger
lake. It was three miles long and a tenth as wide, very
straight, trending north forty degrees west. It lay in
what resembled a small rift valley. Called San Andreas
Lake, it lay in the trace of the fault. Here, actually, was
where the fault was given its name—by Andrew Lawson, in 1895, who thought he was describing a local fea-

ture when in fact it extended more than seven hundred miles. Rock of the fault zone, frequently mashed, erodes easily, and the erosion leaves a groove in the terrain. The groove ran on as far as we could see, and included a second and much longer lake, called Crystal Springs Reservoir. Both lakes were man-made—a term, in this milieu, that women might be pleased to accept. In California, the San Andreas Fault is used as a place to store drinking water. In ditches and pipes the water travels a hundred and fifty miles from Hetch Hetchy Reservoir, in the Sierra Nevada, which lies in a valley adjacent to and equal to Yosemite. In 1913, environmentalists led by John Muir lost America's first great conservation battle when the valley of the Hetch Hetchy was dammed. Above San Andreas Lake, we could see over a flanking ridge and down through a deep pool of sky to San Francisco International Airport, at the edge of the bay. One after another—up through the pool slowly—747s were rising.

The San Andreas and Crystal Springs reservoirs were built before the turn of the century. On April 18, 1906, when the fault-zone surface in northern California was ruptured for nearly three hundred miles, the parallel sides of the reservoirs slid in different directions. The motion here was eight feet. The Pacific side moved north. The rupture went straight up both lakes, but the dams did not break. Nor have they since. They held in March, 1957, when a 5.5 temblor epicentered near Mussel Rock tore open a smaller reservoir. And they held in the Loma Prieta earthquake, of October, 1989.

"There's a seismic gap in here that did not get filled

by that last event," Moores remarked. He meant that in 1989 in this stretch of the San Andreas Fault—well north of the epicenter—the two sides did not move.

The idea of the seismic gap first occurred to the seismologist Akitsune Imamura, in Tokyo, more or less at the time of the great San Francisco earthquake of April, 1906. As he studied Japanese earthquake records, which went back hundreds of years, Imamura arranged them graphically in zones of time and place. Where he found quiescent stretches—unfilled areas of his charts—he could see that they had been temporary, as pressure built to fill them. He could see that Tokyo— for what was then the time being—was in a large quiescent zone. In 1912, he began warning the public that the Tokyo gap was soon to be filled. He said that its size suggested to him a severe shock. Essentially, no one was interested. Imamura repeated his warnings for eleven years. The response remained as empty as the gap. In 1923, a hundred and forty thousand people died as Imamura's gap, in a couple of minutes, closed.

In 1906, when fewer than a million people lived near San Francisco Bay, an estimated three thousand died as a result of the earthquake. The population now exceeds six million, and the much publicized fact that the region is traceried with active faults—that the San Andreas system is not just one trace but a whole family of faults in a stepped and splintering band a great many miles wide—has done nothing to discourage the expanding populace from creating new urban shorelines and new urban skylines and so crowding the faults themselves that the faults' characteristic landforms are obscured beneath tens of thousands of buildings and homes. In addition to troughs and sag ponds, the motion of transform faults produces scarps, scarplets, saddles, notches, kerncols, kernbuts, and squeeze-up blocks. Streambeds are offset. Alluvium is ponded. Undrained depressions form, and parallel ridges and shutter ridges. Springs appear, and oases. Scarce has the geology made these features afresh when earthmovers move in to move them out, preparing the ground for structures even less permanent. One has to travel to remoter parts of the San Andreas Fault to see its full range of geomorphic features. In San Mateo County, the first county south of the city, nearly all such features have been obscured or destroyed by housing since 1945. Greater San Francisco, the most beautiful urban landscape in the United States, just will not be inconvenienced by a system of sibling faults. Less than a year after the major earthquake of 1989, a modest (fourteen-hundred-square-foot) two-bedroom house in the Marina, the most devastated residential district in San Francisco, could be had for five hundred and sixteen thousand dollars, a fall of barely ten per cent from pre-earthquake prices.

Visiting California after the 1906 earthquake, Harry Fielding Reid, of Johns Hopkins University, conceived the theory of elastic rebound, which is also known as the Reid mechanism. It describes the mechanics of fault motion. It preceded by sixty years the larger story of where such motions can lead. In a couple of hundred miles of the San Andreas fault trace, nature's hints to Harry Reid were not faint. He saw offset crop rows, tree lines, and fences. He found tunnels, highways, and bridges misaligned. Reid decided that elastic strain must have accumulated for years in the rock below until a moment came when the strain surpassed the strength of the rock, causing an abrupt slip, which released the stored energy.

Because the slip followed the direction—the strike—of the rupture, the San Andreas was a strike-slip fault. It was also known as a wrench fault. Not until the discovery of plate tectonics would it also be called a transform fault. The magnitude of the slip—the jump—diminishes with distance from the epicenter. In Marin County in 1906, a dirt road was severed where it crossed the Olema Valley at the head of Tomales Bay. It sprang apart twenty feet. That, or something near it, was the earthquake's maximum jump. The epicenter was underwater, not far away. The shaking lasted a full minute in 1906—four times as long as the shaking in the event of October, 1989, which released about one thirty-fifth as much energy.

Tomales Bay, long and narrow, resembles San Andreas Lake, and is also in a trough directly on the fault. Standing on its shore, you are impressed not only by its fjordlike dimensions but even more by the complete dissimilarity of the two sides. A tan cotton sock on one foot and a green wool sock on the other could not represent a greater mismatch. On the east side of Tomales Bay are bald unpopulated hills, straw brown in most seasons, and a scatter of lone oaks. Over the west shore of the bay—above the small riparian towns—is a dark-green vegetated ridge, a comparative jungle, which expresses the geology beneath it. The rock of the east shore is Franciscan mélange, and presents at its surface a typical Coast Range demeanor. The west side is, for the most part, granite. In age, the two formations are millions of years apart, but, more to the point, they are different in provenance as well. The granite on the west side of Tomales Bay, like the granite under the sea off Mussel Rock, broke away from the southern Sierra Nevada and has travelled north along the fault at least three hundred miles, an earthquake at a time.

Similar offsets line the great fault. A Cretaceous quartz monzonite on the east side of the fault in San Bernardino County mirrors a Cretaceous quartz monzonite on the west side of the fault near San Luis Obispo. An Eocene sandstone on the east side of the fault near San Luis Obispo appears to be the same as an Eocene sandstone on the west side of the fault in the Santa Cruz Mountains. The volcanics of Pinnacles National Monument, on the west side of the fault at the latitude of Monterey, evidently broke away from a chemically identical formation that remains on the North American side of the fault some two hundred and fifty miles south. (Wedged into a tight canyon at Pinnacles with his entire family one day, Moores found himself intoning, "Fault, don't move now.") A Cretaceous gabbro on the east side of the fault in Santa Barbara County closely matches a Cretaceous gabbro on the west side of the fault in Mendocino County, three hundred and sixty miles away. Included in the gabbros in both places are bits of rare purple amygdaloidal andesite. On the peninsula south of San Francisco is a piece of a structural basin that seems to have broken away from the San Joaquin Basin, two hundred miles down the fault. In the nineteen-sixties, as the theory of plate tectonics was emerging and all this motion was beginning to be understood as the steady movement of the Pacific and North American plates sliding past each other, a foundation hole was dug for a nuclear power plant exactly in the trough of the San Andreas Fault at Bodega Head, fifty miles up the trace from San Francisco. Half the hot fuel rods would have ended up in the Tropics and the other half in Alaska, but environmentalists halted the project. The Salinian Terrane, sliding past North America with San Diego aboard—and Big Sur and Salinas and Santa Cruz—will, in time, carry Los Angeles to San Francisco. Meanwhile, a part of northern Salinia is the Point Reyes Peninsula—the granite west of Tomales Bay. Looked at from the air, the Point Reyes Peninsula seems about as disjunct from the rest of California as Saudi Arabia is from Africa, and for the same reason: a boundary of lithospheric plates.

Along the San Andreas Fault, the average annual rate of slip is enough to transport something one nautical mile in sixty thousand years. Locally, the jump in 1906 represented roughly two hundred years. In some parts of the fault system, the motion assumes an almost steady creep, but, over all, most of the slip is staccato in time and occurs in elastic rebounds. The strain is essentially constant as the Pacific Plate tugs northwest. In response, earthquakes occur annually in the tens of thousands, most of them below the threshold of human sensitivity. Where the two sides of the fault are most tightly locked, the strain builds highest before it goes. The event of 1906 was what is now known as a large plate-rupturing earthquake. Vertically, the earth broke all the way down to the lower crust. Laterally, it opened the surface like a zipper—from the epicenter northward, where the fault trace for the most part lies just offshore and parallel to the coast, and southward, throwing up what appeared to be a plowed furrow through the rifted hills of Marin, tearing the seafloor where the fault passes west of the Golden Gate, opening the cliff at Mussel Rock, splitting the San Francisco Peninsula, and stopping near San Juan Bautista, east of Monterey Bay. Only once in the historical record has a jump on the San Andreas exceeded the jump of 1906. In 1857, near Tejon Pass outside Los Angeles, the two sides shifted thirty feet.

Kerry Sieh, a San Andreas specialist at Caltech, has dug trenches in numerous places across the fault zone near Los Angeles in order to examine the evidence in the exposed sediments. He has established that twelve great events have occurred on the south-central San Andreas Fault in the past two millennia, with intervals averaging a hundred and forty-five years. The Tejon event of January 9, 1857, is the most recent. One does not have to go to Caltech to add a hundred and forty-five to that.

In 1992, the United States Geological Survey completed a series of studies of the fault segment near San Francisco, and concluded that earthquakes on the order of magnitude of the 1906 event—it has been estimated at 8.3—probably recur every two hundred and fifty years. To the human eye, such a number appears in dim light. In a country where people get up in the dead of night to see what has happened on the Tokyo market, who is worried about two hundred and fifty years? When your complete range of concern begins with your grandparents and stops with your grandchildren, one of your safest bets is elastic rebound. In the prodigious roster of earthquakes on the San Andreas Fault, nearly all of them affect no one. The plates drift, the people with them. Fourteen times a year, an earthquake on the order of the 1989 event near Loma Prieta in the Santa Cruz Mountains occurs somewhere in the world. That might seem to thicken the risk. But not much. In California, only six events have occurred at that level since 1900. So why not move in, spread out, build up, lay back, occupy this incomparable terrain?

It is said that if a cow lies down in California a seismologist will know it. In Iceland, which is seismically one of the most active countries in the world, there are fewer than thirty monitored seismographs. In California, there are seven hundred. Among countries of the world, only Japan has more seismographs than California. To track plate motions on both a large and a local scale, geologists also use very-long-baseline interferometry, in which the arrival times of noise from quasars at widely separated stations on earth are used to measure distances (even very long distances) with an error margin of less than a centimetre. (They can measure the actual distance that Africa and South America move apart in one calendar year.) The seismographs, the V.L.B.I.s, and other devices enabled Lynn Sykes and Stuart Nishenko, of Columbia University's Lamont-

John McPhee

Doherty Earth Observatory, to predict in 1983 that "the segment of the San Andreas fault from opposite San Jose to San Juan Bautista, which ruptured less than 1.5 m in 1906 and which probably also broke in 1838, is calculated to have a moderate to high probability of an earthquake of magnitude 6¾ to 7¼ during the next 20 years." In 1989, after that particular stretch of the San Andreas Fault produced a magnitude 6.9 temblor, television interviewers rolled their eyes toward sound-proof ceilings when geologists told them that predictions could not be more exact than a time frame of twenty years. This was too much for the anchor flukes, who think in airtime. But a ratio of one moment to twenty years actually represented an amazing juxtaposition of human time and geologic time. To predict a major earthquake within twenty years was like shooting out a candle flame at five thousand yards.

Sykes and Nishenko also noted that the segment of the San Andreas Fault between Tejon Pass and the Salton Sea was another place quite likely to experience a major event. In the lower part of that region, east of Los Angeles and San Diego, there has not been a great earthquake—8.0 and larger—in historic time. A loud announcement that such an event may be forth-coming was made by the Joshua Tree, Landers, and Big Bear earthquakes of 1992, which occurred fairly near that part of the San Andreas and effectively increased the stress upon it. The combined power of Landers and Big Bear, which came on the same day, caused oil to flow heavily from mountain seeps a hundred and fifty miles west, further endangering the endangered threespine

[242]

stickleback (a fish). Astonishingly, the Landers and Big Bear earthquakes and one that occurred at Cape Mendocino, all in 1992, rank among the twelve largest earthquakes that have happened in California in the twentieth century. With three earlier and lesser earthquakes, Joshua Tree and Landers form a straight line pointing north. They have broken open a new fault. Like a river seeking a straight path, the San Andreas seems to want to shift direction and go north through the Mojave Desert and up the east side of the Sierra Nevada (a probability mentioned to me in 1978 by the geologist Kenneth Deffeyes). The Landers earthquake, as if to emphasize the significance of the new fault vector, reached 7.5 and would have been extremely destructive had it happened in a setting other than a desert. Even before 1992, the accumulated strain on the nearby San Andreas was thought to be enough to open a plate-shattering rupture two hundred miles long. In 1981, the Federal Emergency Management Agency published a warning that a temblor of 8.3 or better was likely to occur on that part of the San Andreas before the end of the century. The agency said that property losses would amount to roughly twenty billion dollars, large numbers of people would be hospitalized, and as many as fourteen thousand would be dead. In 1982, the California Division of Mines and Geology chimed in with a special publication describing the same putative earthquake as "an event expected to take place during the lifetime of many of the current residents of southern California," and going on to say that "two of the three major aqueduct systems which cross the San Andreas fault will be ruptured and supplies will not be restored for a three- to six-month period."

In the same year, the California Division of Mines and Geology issued a companion publication called "Earthquake Planning Scenario for a Magnitude 8.3 Earthquake on the San Andreas Fault in the San Francisco Bay Area." The scenario predicted that the Bay Bridge would withstand the shaking. So would the Golden Gate. Of a viaduct running through the Marina the scenario said it would "collapse." Of the Nimitz Freeway in Oakland the scenario said, "The hydraulic fills used to construct miles of freeway along the east shore of the Bay in Alameda County may liquefy during heavy shaking, with long sections becoming totally impassable. . . . The elevated section through downtown Oakland is expected to be extensively damaged." The earthquake of the prediction was thirty-five times as intense as the earthquake that actually came.

By the time of these scenarios, the rock offsets along the San Andreas had been explained, and the role of earthquakes at the plate boundary was understood. In 1906, the great earthquake was an unforeseeable Act of God. Now the question was no longer *whether* a great earthquake would happen but *when.* No longer could anyone imagine that when the strain is released it is gone forever. Yet people began referring to a chimeric temblor they called "the big one," as if some disaster of unique magnitude were waiting to happen. California has not assembled on creep. Great earthquakes are all over the geology. A big one will always be in the offing. The big one is plate tectonics.

At one time and another, for the most part with Moores, I have travelled the San Andreas Fault from the base of the Transverse Ranges outside Los Angeles to the rocky coast well north of San Francisco. In clear weather, a pilot with no radio and no instrumentation could easily fly those four hundred miles navigating only by the fault. The trace disappears here and again under wooded highlands, yet the San Andreas by and large is not only evident but also something to see—like the beaten track of a great migration, like a surgical scar on a belly. In the south, where State Route 14 climbs out of Palmdale on its way to Los Angeles, it cuts across the fault zone through two high roadcuts in which Pliocene sediments look like rolled-up magazines, representing not one tectonic event but a whole working series of them, exposed at the height of the action. On the geologic time scale, the zone's continual agitation has been frequent enough to be regarded as continuous, but in the here and now of human time the rift extends quietly northward through serene, appealing country: grasses rich in the fault trough, ridges intimate on the two sides—a world of tight corrals and trim post offices in towns that are named for sag ponds.

Farther north, it loses, for a while, its domestic charm. Almost all water disappears in a desert scene that, for California, is unusually placed. The Carrizo Plain, only forty miles into the Coast Ranges from the ocean at Santa Barbara, closely resembles a south Nevada basin. Between the Caliente Range and the Temblor Range, the San Andreas Fault runs up this flat, unvegetated, linear valley in full exposure of its benches and scarps, its elongate grabens and beheaded channels, its desiccated sag ponds and dry deflected streams. From the air, the fault trace is keloid, virtually organic in its insistence and its creep—north forty degrees west. On the ground, standing on desert pavement in a hot dry wind, you are literally entrenched in the plate boundary. You can see nearly four thousand years of motion in the bed of a single intermittent stream. The bouldery brook, bone dry, is fairly straight as it comes down the slopes of the Temblor Range, but the San Andreas has thrown up a shutter ridge—a sort of sliding wall—that blocks its path. The stream turns ninety degrees right and explores the plate boundary for four hundred and fifty feet before it discovers its offset bed, into which it turns west among cobbles and boulders of Salinian granite.

You pass dead soda ponds, other offset streams. The (gravel) road up the valley is for many miles directly on the fault. Now and again, there's a cattle grid, a herd of antelope, a house trailer, a hardscrabble ranch, a fence stuffed with tumbleweed, a pump in the yard. A daisy wheel turns on a tower. Down in the broken porous fault zone there will always be water, even here.

With more miles north come small adobes, far apart, each with a dish antenna. And with more miles a handsome spread, a green fringe, a prospering ranch with a solid house. The fault runs through the solid house. And why should it not? It runs through Greater San Francisco.

Of the two most direct routes from southern to northern California, always choose the San Andreas Fault. If you have adequate time, it beats the hell out of Interstate 5. Nearly always, some sort of road stays right in the fault zone. Like a water-level route through rough country, the fault is a place to find gentle grades and smooth ground. When the fault makes minor turns, they are nothing compared to the bends of a river. With more distance north, the desert plain yields to hay meadows and then to ever lusher country, until vines are standing in the fault-trace grabens and walnuts climb the creaselike hills. Ground squirrels appear, and then ever larger flocks of magpies, and then cottonwoods, and then oaks in thickening numbers, and velure pastures around horses with nothing to do. In age and rock type, the two sides of the fault are as different as two primary colors. Strewn up the west side are long-transport gabbroic hills and deracinated ranges of exotic granite. Just across the trough is Franciscan mélange—stranger, messier, more interesting to Moores.

Near Parkfield, you cross a bridge over the San Andreas where Cholame Creek runs on the fault. The bridge has been skewed—the east end toward Chihuahua, the west end toward Mt. McKinley. Between Cholame and Parkfield, plate-shattering ruptures have occurred six times since 1857, an average of one every twenty-two years, and the probability that another will occur before 2003 has been reckoned at ninety-eight per cent. Thirty-seven people live in Parkfield. If the population is ever to increase, seismologists will be the first to know it, for the valley here is wired like nowhere else. Parkfield has attracted earthquake-prediction experts because the brief interval time on this segment of the fault suggests that if they monitor this place they may learn something before they die. Also, the Parkfield segment has—in Moores' words—"relatively simple fault geometry." And the last three earthquakes have had a common epicenter and have been of equal magnitude.

An average of one plate-shattering earthquake every twenty-two years works out to forty-five thousand per million years. The last big Parkfield event was in 1966. It broke the surface for eighteen miles. Words on the town's water tower say "Parkfield, Earthquake Capital of the World, Be Here When It Happens." The actual year doesn't matter much. The instrumentation of Parkfield assumes that a shock is imminent. Its purpose is not to confirm the calculated averages but to develop a technology of sensing—within months, days, hours, or minutes—when a shock is coming. Even a minute's warning, or five minutes', or an hour's, let alone a day's, could (in highly populated places) save many lives and much money. Accordingly, the Cholame Valley around Parkfield—between Middle Mountain, to the north, and Gold Mountain, to the south—has been equipped with several million dollars' worth of strain gauges, creepmeters, earth thumpers, laser Geodimeters, tiltmeters, and a couple of dozen seismographs. It is said that the federal spending has converted the community from Parkfield to Porkfield. Some of the seismographs are in holes half a mile deep. Experience suggests that rocks creep a little before they leap. The creepmeters are sensitive to tens of millionths of an inch of creep.

If ever there was a conjectural science, it is earthquake prediction, and as research ramifies, the Tantalean goal recedes. The maximum stress on the San Andreas Fault—the direction of maximum push—turns out to be nearly perpendicular to the directions in which the fault sides move, like a banana peel's horizontal slip when pressure comes upon it from above. A fault that moves in such a manner must be weak enough to slide—must be, in a sense, lubricated. Among other things, the pressure of water in pores of rock in the walls of the fault has been mentioned as a lubricant, and so has the sudden release of gases that may result from shaking. Such mechanisms would tend to randomize earthquakes, diminishing the significance of mounting strains and temporal gaps. Those who practice earthquake prediction will watch almost anything that might contribute to the purpose. A geyser in the Napa Valley inventively named Old Faithful seems to erupt erratically both before and after large earthquakes that occur within a hundred and fifty miles—an observation that is based, however, on records kept for not much more than twenty years. In 1980, the United States Geological Survey began monitoring hydrogen in soils. Two years later, near Coalinga, about twenty miles northeast of Parkfield, the hydrogen in the soil was suddenly fifty times normal. It appeared in bursts, and such bursts became increasingly numerous in April, 1983. In May, 1983, a 6.4 earthquake occurred on a thrust fault under Coalinga. Releases of radon are watched. So are patterns and numbers of microquakes, especially those that are known as the Mogi doughnut. In the mid-nineteen-sixties, a Japanese seismologist noticed on his seismograms that microquakes occurring in the weeks before a major shock sometimes formed a ring around the place that became the epicenter. Mogi's doughnut is a wonderful clue, but—like hydrogen bursts and radon releases—before most major shocks it fails to appear.

People who live in earthquake country will speak of earthquake weather, which they characterize as very balmy, no winds. With prescient animals and fluctuating water wells, the study of earthquake weather is in a category of precursor that has not attracted funds from the National Science Foundation. Some people say that well water goes down in anticipation of a temblor. Some say it goes up. An ability to sense imminent temblors has been ascribed to snakes, turtles, rats, eels, catfish, weasels, birds, bears, and centipedes. Possible clues in animal behavior are taken more seriously in China and

Japan than they are in the United States, although a scientific paper was published in *California Geology* in 1988 evaluating a theory that "when an extraordinarily large number of dogs and cats are reported in the 'Lost and Found' section of the *San Jose Mercury News,* the probability of an earthquake striking the area increases significantly."

Earthquake prediction has taken long steps forward on the insights of plate tectonics but has also, on occasion, overstepped. Until instrumentation is reliably able to chart a developing temblor, predictors obviously have a moral responsibility to present their calculations shy of the specific. The mathematical equivalent of a forked stick will produce such absurdities as the large earthquake that did not occur as predicted in New Madrid, Missouri, on December 2, 1990. A U.S.G.S. geologist and a physicist in the United States Bureau of Mines whose research included (among other things) the study of rocks cracking in a lab predicted three great earthquakes for specific dates in the summer of 1981, to take place in the ocean floor near Lima. The largest—9.9—was to be twenty times as powerful as any earthquake ever recorded in the world. A few hundred thousand Peruvians were informed that they would die. Nothing happened.

If you set stakes in a straight line across a valley glacier and come back a year later, you will see the curving manner in which the stakes have moved. If you drive fence posts in a straight line across the San Andreas Fault and come back a year later, almost certainly you will see a straight line of fence posts—unless your fence is in the hundred miles north of the Cholame Valley. There the line will be offset slightly, no more than an inch or two. Another year, and it will have moved a little more; and a year after that a little more; and so forth. In its seven hundred and forty miles of interplate abrasion, the San Andreas Fault is locally idiosyncratic, but nowhere more so than here in the Central Creeping Zone. Trees move, streams are bent, sag ponds sag. In road asphalt, echelon fractures develop. Slivers drop as minigrabens. Scarplets rise. The fault is very straight through the Central Creeping Zone. It consists, however, of short (two to six miles), stepped, parallel traces, like the marks made on ice by a skater. Landslides occur frequently in the Central Creeping Zone, obscuring the fresh signatures of the creep.

"The creep is relatively continuous for a hundred and seventy kilometres here and seems to account for nearly all of the movement," Moores remarked. "Creep is rare. Most fault movement is punctuated. The creep produces numerous small earthquakes. There are actual 'creep events,' wherein as much as five hundred metres of the fault zone will experience propagating creep in one hour." There were many oaks and few people living in the creep zone. The outcrops on the Pacific side of the fault sparkled with feldspar and mica—the granitic basement of the Gabilan Range. More than three thousand feet in elevation and close against the fault trace, the Gabilan Range creeps, too.

Jumping and creeping, the San Andreas Fault's average annual motion for a number of millions of years has been thirty-five millimeters. The figure lags significantly behind the motion of the Pacific Plate, whose travels, relative to North America, go a third again as fast. In the early days of plate tectonics, this incongruous difference was discovered after the annual motion of the Pacific Plate was elsewhere determined. The volcanic flows that crossed the San Andreas and were severed by the fault had not been carried apart at anything approaching the rate of Pacific motion. This became known as the San Andreas Discrepancy. If the Pacific Plate was moving so much faster than the great transform fault as its eastern edge, the rest of the motion had to be taken up somewhere. Movements along the many additional faults in the San Andreas family were not enough to account for it. Other motions in the boundary region were obviously making up the difference.

With the development of hot-spot theory (wherein places like Hawaii are seen as stationary and deeply derived volcanic penetrations of the moving plates) and of other refinements of data on vectors in the lithosphere, the history of the Pacific Plate became clearer. About three and a half million years ago, in the Pliocene epoch, the direction in which it was moving changed about eleven degrees to the east. Why this happened is the subject of much debate and many papers, but if you look at the Hawaiian Hot Spot stitching the story into the plate, you see, at least, that it did happen: there is an eleven-degree bend at Pliocene Oahu.

The Pacific Plate, among present plates the world's largest, underlies about two-thirds of the Pacific Ocean. North-south, it is about nine thousand miles long, and, east-west, it is about eight thousand miles wide. What could cause it to turn? Various events that occurred roughly three and a half million years ago along the Pacific Plate margins have been nominated as the cause. For example, the Ontong-Java Plateau, an immense basaltic mass in the southwest Pacific Ocean, collided with the Solomon Islands, reversing a subduction zone (it is claimed) and jamming a huge slab of the Pacific Plate under the North Fiji Plateau. The slab broke off. Suddenly released from the terrific drag on its southwest corner, the rest of the great northbound plate turned eleven degrees to the northeast. A number of coincidental collisions along the plate's western margin may have contributed to the change in vector. Additional impetus may have been provided by the subduction of a defunct spreading center at the north end of the plate. The extra weight of the spreading center, descending, may have tugged at the plate and given it clockwise torque. Whatever the cause, it's not easy to imagine a vehicle that weighs three hundred and forty-five quadrillion tons suddenly swerving to the right, but evidently that is what it did.

The tectonic effect on North America was something like the deformation that results when two automobiles sideswipe. Between the Pacific and North American plates, the basic motion along the San Andreas Fault remained strike-slip and parallel. But as the Pacific Plate sort of jammed its shoulder against most of California a component of compression was added. This resulted in thrust faults and accompanying folds—anticlines and synclines. (Petroleum migrated into the anticlines, rose into their domes, and was trapped.)

Earlier—about five million years before the present—the ocean spreading center known as the East Pacific Rise had propagated into North America at the Tropic of Cancer, splitting Baja off the rest of the continent, and initiating the opening of the Gulf of California. (By coincidence, the walls of the Red Sea, which much resembles the Gulf of California, parted at the same time.) The splitting off of Baja was accompanied by very strong northward compression, which raised, among other things, the Transverse Ranges above Los Angeles, at the great bend of the San Andreas. That the Transverse Ranges were rising compressionally had been obvious to geologists long before plate tectonics identified the source of the compression. But not until

the late nineteen-eighties did they come to see that compression as well as strike-slip motion accompanies the great fault throughout its length, as a result of the slight shift in the direction of the Pacific Plate three and a half million years ago. All these compressional aspects taken together—anticlines, synclines, and thrust faults in a wide swath from one end of California almost to the other—account for some of the missing motion in the San Andreas Discrepancy. The Los Angeles Basin alone has been squeezed about a centimetre a year for two million two hundred thousand years. The sites of Laguna Beach and Pasadena are fourteen miles closer together than they were 2.2 million years ago. This has happened an earthquake at a time. For example, both the Whittier Narrows earthquake of 1987 and the Northridge earthquake of 1994 lessened the breadth of the Santa Monica Mountains and raised the ridgeline.

The Whittier Narrows hypocenter was in a deeply buried fault in a young anticline. Such faults tend to develop about ten miles down and gradually move toward the surface. Northward for five hundred miles, young anticlines on the east side of the San Andreas Fault are similar in nature—the products of deep successive earthquakes. Most are recently discovered, and many more, presumably, remain unknown. They make very acute angles with the fault, like the wake of a narrow boat. When a temblor goes off like a hidden grenade, geologists often have not suspected the existence of the fault that has moved. The 6.4 earthquake at Coalinga in 1983 was that kind of surprise. It increased the elevation of the ridge above it by more than two feet.

In 1892, a pair of enigmatic earthquakes shook Winters, which is near Davis, in the Great Central Valley. Evidently, the earthquakes occurred on the same sort of blind thrust that is under Coalinga, but the Winters thrust is of particular interest, because it is east of the Coast Ranges and fifty miles from the San Andreas. Yet it is apparently a product of the newly discovered folding and faulting that everywhere shadow the great fault. The Central Valley of California is about the last place in the world where virtually any geologist would look for an Appalachian-style fold-and-thrust belt. Without shame, Moores sketches one on a map of California; it goes up the west side of the valley almost all the way from the Tehachapi Range to Red Bluff and reaches eastward as far as Davis. He and his Davis colleague Jeff Unruh have been out looking for tectonic folds in the surreally flat country surrounding the university. This is a game of buff even beyond the heightened senses of the blind. They have found an anticline—an arch with limbs spread wide for many miles and a summit twenty-five feet high. They call it the Davis Anticline. It is a part of what Moores likes to describe as "the Davis campus fold-and-thrust belt." He is having fun, but the folds are not fictions. The anticline at Davis has developed in the past hundred thousand years. It is rising ten times as fast as the Alps.

On perhaps the weirdest geologic field trip I have ever been invited to observe, he and Unruh went out one day looking for nascent mountains in the calm-water flatness of the valley. There were extremely subtle differentiations. Moores said, "We are looking here

on the surface for something that is happening five kilometres down—blind thrusts. Compressional stress extends to the center of the valley."

"Topography doesn't happen for nothing," Unruh
said. "Soil scientists have long recognized that these valley
rises are tectonic uplifts. Soils are darker in basinal areas.
There's a fault-propagation fold in this part of the valley."

Moores later wrote to me:

We continue to gather evidence. We have seen two seismic profiles that show a horizontal reflection, presumably a
fault, that extends all the way from the Coast Ranges to the
Sacramento River. Jeff has been working at stream gradients.
The rationale is that where there is a sharp change in gradient on a flood plain there is a reason, and the reason here is
uplift. The analysis fits the two areas of acknowledged uplift
west of Davis pretty well, and seems to indicate a new north-
trending zone of uplift that goes right through Davis itself.
Maybe there was a reason why the Patwin Indians selected
this particular spot on the banks of Putah Creek for this village, after all. It was a high spot in a swamp, and it was high
because it is coming up!

The compressive tectonism associated with the
plate boundary contributes to the total relative plate
motion, but not much: the over-all average is less than a
centimetre a year. And that does not nearly close the
numerical gap. Surprisingly, the rest of the missing motion seems to come from the Basin and Range, the
country between Reno and Salt Lake City, wherein the
earth's crust has been stretching out and breaking into

blocks, which float on the mantle as mountains. The stretching has increased the width of the region by sixty miles in a few million years. Very-long-baseline interferometry has shown that the Basin and Range is spreading about ten millimetres a year in a direction west-northwest. This supplies enough of the total plate-boundary motion between the Gulf of California and Cape Mendocino to make up the difference in the San Andreas Discrepancy. If some Pacific Plate motion is coming from Utah, Utah is a part of the plate boundary.

The westernmost range of the Basin and Range Province is the Sierra Nevada, which has risen on a normal fault that runs along the eastern base of the mountains. The fault has experienced enough earthquakes to give the mountains their exceptional altitude. The most recent great earthquake there was in 1872. In a few seconds, the mountain range went up three feet. In the same few seconds, the Sierra Nevada also moved north-northwest twenty feet. That would help to fill in anybody's discrepancy.

Perhaps a sixth of the total motion between the plates is contributed by the other faults in the San Andreas family. Each is strike-slip, active right-lateral—that is, viewed from one side of the fault, the other side appears to have gone to the right.

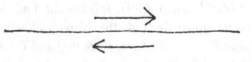

In a general way, you can demonstrate their relationship to one another with a deck of cards. Hold the deck, side up, between the palms of your hands, and slide the hands, pulling the right side toward you, pushing the left side away, and keeping pressure on the deck. The cards will respond by slipping, sticking, locking, sliding. Some may slide more than others. There may even develop a primary break. In any case, the fifty-one slips between the cards are, as in California, a family of right-lateral strike-slip faults. If one has moved more than the others, in effect you may have cut the cards, and you could call that cut the San Andreas Fault. But all the cards, to varying extents, have contributed slip to the total motion.

Moores believes that the plate-vector change three and a half million years ago is what probably created so large a grouping of boundary faults. Parallel and sub-parallel to the San Andreas, they have been likened to the tributaries of a river or the branches of a tree. But they are not dendritic. Often they do not conjoin. They are more like the checks that appear in dry timber. They all trend northwest. Many of them have varying local names, because the field geologists who did the naming, as much as a hundred years ago, did not suspect their continuity. The actively slipping Green Valley Fault, which intersects I-80 at Cordelia, in the eastern Coast Ranges, continues to the south as the Concord Fault. I-680, branching off, follows the fault and stays right on it. Down through the Napa Valley and under San Pablo Bay and through Berkeley and Oakland and Hayward and farther south than San Jose runs a continuous fault that is in segments named Healdsburg, Rodgers Creek, and Hayward. Portentous microquakes on the Rodgers Creek Fault have suggested to the United States Geological Survey the possibility of an earthquake equal in intensity to the 1989 wrench on the San Andreas near Loma Prieta.

The Calaveras Fault runs close to the Hayward Fault and extends somewhat farther south, like the Sargent Fault, the Wildcat Fault, the Busch Ranch Fault. The Antioch Fault is in the Great Central Valley. In seven hundred miles of splintery faults, the ones I have mentioned are all in the San Francisco Bay Area east of the San Andreas. West of it—on the San Francisco Peninsula or under the ocean—are the Pilarcitos Fault, the La Honda Fault, the Hosgri Fault, the San Gregorio Fault. The San Gregorio Fault extends from San Francisco to Big Sur, south of Monterey. Its longest historical jump is thirteen feet. It has produced great or major earthquakes on an average of once every three hundred years—nothing to be concerned about, unless it is your year. In the San Gregorio fault zone, San Mateo County has forty-acre zoning.

A comparable cross-sectional anatomy of the San Andreas system could be described for any latitude from Cape Mendocino to the Salton Sea, with a long list of names of contributive strike-slip faults. Within the system, the twentieth-century earthquake second in severity occurred in 1952 on the White Wolf Fault, near Bakersfield. In southern California, the belt is as much as a hundred and fifty miles across—three times as wide as it is at San Francisco, and a good deal more complex.

The Hayward Fault alone has contributed more than a hundred miles of offset. Running southeast from

San Pablo Bay, north of San Francisco, to the latitude of Santa Cruz, it disappears near Gilroy, not far from the San Andreas Fault at San Juan Bautista. In many places, such as Berkeley, the Hayward Fault has Jurassic rock of the Franciscan mélange on one side of it and Cretaceous rock of the Great Valley Sequence sort of dredged up on the other side. A return specialist in the football stadium going eighty yards through a broken field will gain or lose about fifty million years. In a split, unpredictable second, he can be tossed out of bounds by a shift of the sod beneath him. The Hayward Fault runs not only through Memorial Stadium but also through or very near the Alameda County hospital, the San Leandro hospital, and California State University, Hayward. The Hayward Fault also ran through the California School for the Deaf and Blind, but the State became nervous, moved the school to another site, and then filled up its old dorms with Berkeley undergraduates.

The Hayward Fault separates the Cretaceous Berkeley Hills from the Jurassic university campus and the Holocene alluvium of the flat ground near San Francisco Bay. To no small extent, the Hayward Fault has created the Berkeley Hills, which are an obvious fault scarp, the result of a vertical component in an otherwise strike-slip motion. The change is abrupt from gentle slope to steep escarpment because the fault is so active and the hills are so young. Large earthquakes occurred on the Hayward Fault in 1836 and 1868—not the sort of information that is likely to plumb the tilt in a laid-back sophomore at Berkeley. However, a

U.S.G.S. Miscellaneous Field Studies map of predicted
maximum ground-shaking from large earthquakes on
the San Andreas Fault and the Hayward Fault shows
three areas of A-level intensity—characterized as "very
violent"—for a Hayward earthquake: the environs of
lower University Avenue in Berkeley; the blocks just
east of Lake Merritt in downtown Oakland; and the
Warren Freeway in Piedmont. The Warren Freeway
uses the Hayward Fault in the way that water uses a
riverbed.

The Geological Survey sees a sixty-seven-percent
chance of another major San Francisco Bay Area earthquake on either the San Andreas or the Hayward Fault
before the year 2020, with probabilities leaning toward
the Hayward, because a jump there is so long overdue.
If it should equal the intensity of several nineteenth-
century shocks, the new one—according to estimates
by the Federal Emergency Management Agency and
the California Division of Mines and Geology—would
result in as many as four thousand five hundred deaths,
a hundred and thirty-five thousand injuries, and forty
billion dollars' worth of damage. The Geological Survey
adds a comment to these possibilities: the center of San
Francisco "is as close to the Hayward Fault as it is to
the San Andreas Fault."

San Lorenzo Creek, coming out of the hills into
the outskirts of Hayward, bends sharply right when it
hits the fault, flows northwest on the fault for more
than a mile, and then takes a left and heads southwest
to the bay. Hayward is about fifteen miles down the
trace from Berkeley, and if ever there was a type local-

ity to define the geological meaning of "type locality" Hayward is the place. Rarely is a fault zone so sharply drawn. Through part of the town runs a steep emphatic bench, resembling a sloped medieval wall, that is the product of the fault beside it. On D Street between Mission and Main, the curb, running east, makes a right-lateral bend to the south, and then continues east. When Moores and I were on D Street not long ago, a pressure ridge had buckled the sidewalk. The fault went through the service department of Boulevard Buick. There were fresh patches in the sidewalk outside. Between C and D, a building that had long housed the municipal government—and had been designed and constructed as the Hayward City Hall— stood precisely on the fault, which was tearing the building apart as if it were a tuft of cotton. Tiles and plaster had showered the bureaucrats until they fled. The bureaucracy included a department established to deal with emergencies confronting the city. At 934 C Street was a store that declared itself to be the "Hayward Sewing Center, Alterations." The alterations included a canted curb, a pressure ridge in the sidewalk, and a wall in the act of bulging toward Mexico City. There was a bend in the long wall of the Action Signs Building, at 22534 Mission. The fault had offset the Spoiled Brat Parking Lot. In a municipal parking lot nearby, the lines of meters took right-lateral bends and then resumed their original direction. Robert's School of Karate—in motion relative to the antique shop next door—had recently moved south an inch. A sign in the window said "KARATE KENPO GUNG-FU BOXING—

The Only School in Hayward Teaching Street Fighting." Sidewalks were patched on every east-west street. The occupied house at 923 Hotel Avenue had been so torqued by the fault that one of its walls was concave. Looming over it—fifty feet high—was the Hayward escarpment.

In the country south of Hayward, we saw numerous streams that came down through the hills, made a right turn at the fault line, and then, after a bit, rediscovered their offset beds. We saw a deep ravine bent like a crochet hook. Like the San Andreas, the Hayward Fault seems to be stuck hard in some places (Berkeley) and in others (mainly in the south) to be creeping. A culvert curves like macaroni. A water tunnel begins to leak. Railroad tracks move. In Fremont, Moores and I climbed over some walls and fences in order to get up onto the ballast and squint down the rails of the Union Pacific, which, where it crossed the Hayward Fault, was bent into an echelon of kinks.

In 1986, a small earthquake on the Quien Sabe Fault, a deeply hidden blind thrust near the south end of the Hayward Fault, sent out elastic waves that shook open a twenty-thousand-gallon vat of cabernet sauvignon in an Almaden winery twenty-five miles from the epicenter. The vat was thirty feet high. A thunderous winefall flooded an office, broke out of the building, and poured down the road. This seems to have done it for Almaden. The Quien Sabe Fault? Like almost everybody else, Almaden had never heard of it, but the Quien Sabe Fault added insult to chronic injury. Ever since the winery was built, it had slowly been coming apart. Whereas the little Quien Sabe Fault was twenty-five miles away, the San Andreas Fault happened to run right through the building. The road outside was called Cienega, meaning "swamp," as in reed-filled sag pond, an example of which was beside the winery. *Almaden* is a Spanish geological term meaning "mine," and this location—about twelve miles south of Hollister, in the Central Creeping Zone—was no place to mine wine. Nevertheless, like so many parts of the San Andreas trace, it was an intimate and lovely valley, full of walnut orchards and olive groves and horse pastures and signs of anxious warning: "CAUTION: CHILDREN WALKING TO SCHOOL."

This was the place where slow tectonic creep, also called aseismic slip, was first observed. The winery stands quiet now under Atlas cedars. Almaden has shut it down. Isabel Valenzuela, a caretaker who had left Mexico two weeks before, gave Moores and me a tour in Spanish. In the gloom of the great space, casks were ranked, and not a few were broken—cracked, like immense standing eggs. Through the floor a wide crack ran from one end to the other of the long rectangular building. Outside was a bronze plaque mounted on a freestanding wall of unreinforced masonry. It bore the words "San Andreas Fault has been designated a Registered Natural Landmark."

I am from the Northeast and have never felt a destructive earthquake. The closest one to my home came in 1980, when a temblor centered in Cheesequake, New Jersey, caused no damage. My wife and I were in San Francisco in mid-October, 1989, and departed by train from Oakland, heading north. A niece in Berkeley drove us to the station and had difficulty finding it, in dark streets among warehouses on low flat bayfill ground. Up one street and down the next she searched— back and forth, around and under the Nimitz Freeway. The earthquake came more than a hundred hours after we left—an inexpressibly short span on the geologic scale, an irrelevant long one on ours. I wish I could say that I felt in my neuroplasm that it was time to go, but I was not a missing cat.

A few weeks later, when I was back in California, Moores and I were approaching the Santa Cruz Mountains from the south, and we stopped off at the Spanish mission in San Juan Bautista. Where scouts discovered two wells of water only twenty paces apart, the mission was established in 1797. Down the middle of a broad plain ran a sharply defined escarpment about fifty feet high, like the one in Hayward. It was a single long step, steeply inclined, like a grandstand. A modern geologist seeing such a break with springs or shallow wells beside it would be wondering not "Why is it there?" but "How frequent is the slippage upon it?" Moores remarked, "A fault is a good place for a well. It's brecciated. There's lots of porosity. Aquifers on either side are truncated, and spill into the fault." The Franciscans built their mission on the top edge, the brink—not near, but on, the San Andreas. By October, 1800, they had begun or completed eight buildings, of adobe roofed with sedge, when earthquake after earthquake—as many as six shocks a day—brought much of the mission down.

The cool cloisters look immovable, the chapel

looks repaired. Its plans (unique among the missions) called for three aisles divided by columns, but the colonnades were finished as walls. The view from the scarp is much the same: the fault is about all that has moved. In April, 1906, when San Juan Bautista was the southern end of the plate-shattering rupture, there was some destruction. Now, in 1989, there was no sign of the shaking of a few weeks before. Damage had been essentially nil in San Juan Bautista and everywhere in the fault zone to the south.

The fault scarp beside the chapel was actually used as a grandstand for dog races on a dirt track below, but the fans lost interest, and vines have grown over the seats. Beneath the bells of the mission, we sat in the bleachers and talked among the crawling vines, right on the fault, looking east toward distant mountains across the unaltered plain. Like almost all topography, it seemed to be immutable.

"People look upon the natural world as if all motions of the past had set the stage for us and were now frozen," Moores remarked. "They look out on a scene like this and think, It was all made for us—even if the San Andreas Fault is at their feet. To imagine that turmoil is in the past and somehow we are now in a more stable time seems to be a psychological need. Leonardo Seeber, of Lamont-Doherty, referred to it as the principle of least astonishment. As we have seen this fall, the time we're in is just as active as the past. The time between events is long only with respect to a human lifetime."

(In a 1983 paper called "Large Scale Thin-Skin

Tectonics," Seeber addressed the possibility that crustal deformations have occurred on an areal and cataclysmic scale never imagined or described. "Our direct view of geologic phenomena has been severely limited by the relatively short span of history and by the relatively small vertical extent of outcrops," he wrote. "In many respects we only have a two-dimensional snapshot view of the geologic process. Moreover, the interpretation of geologic data was probably influenced by the psychologic need to view the earth as a stable environment. Manifestations of current tectonism were often perceived as the last gasps of a geologically active past. Thus, subjected to the principle of least astonishment, geologic science has always tended to adopt the most static interpretation allowed by the data.")

Earthquakes in the six-to-seven range occurred at least once a decade in the San Francisco Bay Area between 1850 and 1906. Afterward, that segment of the San Andreas Fault was essentially quiet for more than eighty years, a not very significant exception being the 5.3 event near Mussel Rock in 1957. While strain accumulated along the San Francisco Peninsula, whole lifetimes passed, so the principle of least astonishment, which works to a fare-thee-well in a place like New York, seemed to be working even here. In recent days, of course, the newspapers had been full of comment suggesting that least astonishment was no longer a principle in the Bay Area. Withal, there was an undercurrent of implication that it had not died and would come back.

Jerry Carroll, in the *San Francisco Chronicle*:

There is no greater betrayal than when the earth defaults on the understanding that it stay still under foot while we go about the business of life, which is full enough of perils as it is.

Stephanie Salter, *San Francisco Examiner*:

A traumatic experience . . . started in the depths of the earth and wreaked damage all the way to the depths of the psyche. . . . Or maybe the truth is, earthquake time is the most real time of all, a time when all the bull ceases and the preciousness of life is understood most acutely.

Herb Caen, the venerable columnist of the *San Francisco Chronicle,* who had seen his share of accumulating strain:

[This is] a headstrong, careless city dancing forever on the edge of disaster. . . . We realize afresh the joys and dangers of living here, and we reaffirm our belief that it is worth the gamble, however great. . . . We have been validated as San Franciscans.

A few miles up the trace, we looked across the Pajaro Valley at the high notch where the fault slices into the southern extremities of the Santa Cruz Mountains. The east side was Oligocene shale, Moores said, and on the west side was a quartz diorite of the mid-Cretaceous. The two formations differed by at least sixty million years and by who knows how many miles of sliding offset. It was as if an apple and a pear had

both been vertically sliced in half, and two of the differing halves had been placed together to make the mountains. The lookoff where we stood was at the northern
end of the Gabilan Range. The Pajaro River, narrow
and slow, ran westward toward the ocean through a
topographic gap that punctured the mountains. Indians
of the eighteenth century informed the arriving Spanish
that the small stream in that huge gap had not always
been so modest, that it had once been the outlet of the
interior rivers, also draining the bays—the role now
played by the Golden Gate. The Indians were right,
Moores said. Never mind that they may not have
suspected that the whole of the coastal country was
moving northwest, occluding what lay to the east. Geologists have described the Pajaro Valley as "one of the
most seismically active regions in the coterminous
United States." On April 18, 1906, a freight train crossing the valley was thrown off the tracks.

The San Andreas Fault was exceptionally smelly
where it crossed the Pajaro River and went into the
Santa Cruz Mountains. Highway 129 follows the right
bank there. From the fault zone—a landslide of sedimentary hash about a hundred and twenty yards
wide—a dozen sulphur springs were pouring. Galva-
nized pipes had been driven into the springs, causing
the water to spout onto the roadside and color it yellow-
cream.

Just to the west was Watsonville, which looked like
a French battle town in 1944. The eighty-six-year-old
St. Patrick's Church, until recently a steepled brick
structure with four spires, stood in its own red scree.

That it stood at all was remarkable. Its crosses were aslant, its buttresses denuded, its brick gone in swatches from the walls. Like cartoon lightning, jagged cracks descended through the brick that remained. Two spires were stretched out in the church parking lot. Where Highway 1 crossed Struve Slough on concrete columns, the columns had punctured the pavement and now protruded upward like standing stones. Wide acreages in the town center had been bulldozed bare and brown. Ford's Department Store (1851) was totally destroyed. Countless buildings were shuttered with plywood or wrapped in chain-link fencing. There were fissures in cement-block walls. Two hundred and fifty houses were off their foundations, many of them crushed like foam cups. There were tents. There were cellar holes where houses had been. In a single long moment at the edge of town, a million apples had fallen to the ground.

California building codes that involve seismic requirements were first written in 1933. They covered school buildings and nothing else. San Francisco did not extend such codes to other structures until the late nineteen-forties, when they appeared in the laws of virtually all communities around the bays and of many around the state. While most buildings are still "precode," what is most remarkable is how effective the codes have been. In the 1989 Loma Prieta earthquake, sixty-two people died. In an earthquake of similar magnitude in Armenia in 1988, fifty-five thousand died. In Mexico City in 1985, ten thousand died. In the Iranian earthquake of 1990, fifty thousand died. The difference may lie partly in luck, in site, in relative intensity, but largely it lies in building codes, and the required or suggested strengthening of existing structures. Certain vulnerabilities notwithstanding, California seems to know what it is up against, and what to try to do about it. Never mind that in October, 1989, twenty-one thousand homes and commercial buildings were cracked, crumpled, or destroyed, and nature's invoice for a few moments of shaking was six billion dollars.

During the previous summer, there had been a 5.2 earthquake in the Santa Cruz Mountains, and a 5.1 quake the year before. These could be looked upon as precursors, but precursors never become such until a large jump follows, and are therefore useless as warnings. A score of 5.2 on the Richter scale is made by an earthquake with three hundred times less energy than the one that shattered Watsonville. Richter was a professor at Caltech. His scale, devised in the nineteen-thirties, is understood by professors at Caltech and a percentage of the rest of the population too small to be expressible as a number. Another professor at Caltech in Richter's time—and someone who manifestly understood the principles involved—was Beno Gutenberg, who provided the data from which the scale was made. The data applied only to southern California; subsequently, Gutenberg and Richter jointly developed the worldwide scale. Gutenberg did not see or hear well and was understandably reluctant to deal with reporters. He generally asked his young colleague Charles F. Richter to explain the scale to them. Since I have no idea how the scale works, let me say only that it is a mathematically derived combination of three scales parallel to one another: a magnitude scale flanked by scales of amplitude and distance. (Amplitude is the height of the mark an earthquake produces on a seismogram.) Where a line drawn between amplitude and distance crosses the central scale, it registers magnitude. With each rising integer on the magnitude scale, an earthquake's waves have ten times as much amplitude and thirty times as much energy. Richter always insisted that it was the Gutenberg-Richter scale.

There is a swerve in the San Andreas Fault where it moves through the Santa Cruz Mountains. It bends a little and then straightens again, like the track of a tire that was turned to avoid an animal. Because deviations in transform faults retard the sliding and help strain to build, the most pronounced ones are known as tectonic knots, or great asperities, or prominent restraining bends. The two greatest known earthquakes on the fault occurred at or close to prominent restraining bends. The little jog in the Santa Cruz Mountains is a modest asperity, but enough to tighten the lock. As the strain rises through the years, the scales of geologic time and human time draw ever closer, until they coincide. An earthquake is not felt everywhere at once. It travels in every direction—up, down, and sideways—from its place and moment of beginning. In this example, the precise moment is in the sixteenth second of the fifth minute after five in the afternoon, as the scales touch and the tectonic knot lets go.

The epicenter is in the Forest of Nisene Marks, a few hundred yards from Trout Creek Gulch, five miles north of Monterey Bay. The most conspicuous nearby landmark is the mountain called Loma Prieta. In a curving small road in the gulch are closed gates and speed bumps. PRIVATE PROPERTY, KEEP OUT. This is steep terrain—roughed up, but to a greater extent serene. Under the redwoods are glades of maidenhair. There are fields of pampas grass, stands of tan madrone. A house worth two million dollars is under construction, and construction will continue when this is over. BEWARE OF DOG.

Motion occurs fifty-nine thousand eight hundred feet down—the deepest hypocenter ever recorded on the San Andreas Fault. No drill hole made anywhere on earth for any purpose has reached so far. On the San Andreas, no earthquake is ever likely to reach deeper. Below sixty thousand feet, the rock is no longer brittle.

The epicenter, the point at the surface directly above the hypocenter, is four miles from the fault trace. Some geologists will wonder if the motion occurred in a blind thrust, but in the Santa Cruz Mountains the two sides of the San Andreas Fault are not vertical. The Pacific wall leans against the North American wall at about the angle of a ladder.

For seven to ten seconds, the deep rockfaces slide. The maximum jump is more than seven feet. Northwest and southeast, the slip propagates an aggregate twenty-five miles. This is not an especially large event. It is nothing like a plate-rupturing earthquake. Its upward motion stops twenty thousand feet below the surface. Even so, the slippage plane—where the two great slanting faces have moved against each other—is an irregular oval of nearly two hundred square miles. The released strain turns into waves, and they develop half a megaton of energy. Which is serious enough. In California argot, this is not a tickler—it's a slammer.

The pressure waves spread upward and outward about three and a half miles a second, expanding, compressing, expanding, compressing the crystal structures in the rock. The shear waves that follow are somewhat slower. Slower still (about two miles a second) are the surface waves: Rayleigh waves, in particle motion like a rolling sea, and Love waves, advancing like snakes. Wherever things shake, the shaking will consist of all these waves. Half a minute will pass before the light towers move at Candlestick Park. Meanwhile, dogs are barking in Trout Creek Gulch. Car alarms and house alarms are screaming. If, somehow, you could hear all such alarms coming on throughout the region, you could hear the spread of the earthquake. The redwoods are swaying. Some snap like asparagus. The restraining bend has forced the rock to rise. Here, west of the fault trace, the terrain has suddenly been elevated a foot and a half—a punch delivered from below. For some reason, it is felt most on the highest ground.

On Summit Road, near the Loma Prieta School, a man goes up in the air like a diver off a board. He lands on his head. Another man is thrown sideways through a picture window. A built-in oven leaves its niche and shoots across a kitchen. A refrigerator walks, bounces off a wall, and returns to its accustomed place. As Pearl

Lake's seven-room house goes off its foundation, she stumbles in her kitchen and falls to the wooden floor. In 1906, the same house went off the same foundation. Her parents had moved in the day before. Lake lives alone and raises prunes. Ryan Moore, in bed under the covers, is still under the covers after his house travels a hundred feet and ends up in ruins around him.

People will come to think of this earthquake as an event that happened in San Francisco. But only from Watsonville to Santa Cruz—here in the region of the restraining bend, at least sixty miles south of the city—will the general intensity prove comparable to 1906. In this region are almost no freeway overpasses, major bridges, or exceptionally tall buildings. Along the narrow highland roads, innumerable houses are suddenly stoop-shouldered, atwist, bestrewn with splinters of wood and glass, even new ones "built to code." Because the movement on the fault occurs only at great depth, the surface is an enigma of weird random cracks. Few and incongruous, they will not contribute to the geologic record. If earthquakes like Loma Prieta are illegible, how many of them took place through the ages before the arrival of seismographs, and what does that do to geologists' frequency calculations?

Driveways are breaking like crushed shells. Through woods and fields, a ripping fissure as big as an arroyo crosses Morrill Road. Along Summit Road, a crack three feet wide, seven feet deep, and seventeen hundred feet long runs among houses and misses them all. Roads burst open as if they were being strafed. Humps rise. Double yellow lines are making left-lateral jumps.

Cracks, fissures, fence posts are jumping left as well. What is going on? The San Andreas is the classic right-lateral fault. Is country going south that should be going north? Is plate tectonics going backward? Geologists will figure out an explanation. With their four-dimensional minds, and in their interdisciplinary ultraverbal way, geologists can wiggle out of almost anything. They will say that while the fault motion far below is absolutely right lateral, blocks of rock overhead are rotating like ball bearings. If you look down on a field of circles that are all turning clockwise, you will see what the geologists mean.

Between one circle and the next, the movement everywhere is left lateral. But the movement of the field as a whole is right lateral. The explanation has legerdemain. Harry Houdini had legerdemain when he got out of his ropes, chains, and handcuffs at the bottom of the Detroit River.

All compression resulting from the bend is highest near the bend, and the compression is called the Santa Cruz Mountains. Loma Prieta, near four thousand feet, is the highest peak. The words mean Hill Dark. This translation will gain in the shaking, and appear in the media as Dark Rolling Mountain.

At the University of California, Santa Cruz, three first-year students from the East Coast sit under redwoods on the forest campus. As the shock waves reach them and the trees whip overhead, the three students leap up and spontaneously dance and shout in a ring. Near the edge of town, a corral disintegrates, horses run onto a highway, a light truck crashes into them and the driver is killed. Bicyclists are falling to the streets and automobiles are bouncing. Santa Cruz has been recovering from severe economic depression, in large part through the success of the Pacific Garden Mall, six blocks of old unreinforced brick buildings lately turned into boutiques. The buildings are contiguous and are of different heights. As the shock waves reach them, the buildings react with differing periods of vibration and knock each other down. Twenty-one buildings collapse. Higher ones fall into lower ones like nesting boxes. Ten people die. The Hotel Metropol, seventy years old, crashes through the ceiling of the department store below it. The Pacific Garden Mall is on very-young-floodplain river silts that amplify the shaking—as the same deposits did in 1906.

Landslides are moving away from the epicenter in synchrony with the car alarms. As if from explosions, brown clouds rise into the air. A hundred and eighty-five acres go in one block slide, dozens of houses included. Hollister's clock tower falls. Coastal bluffs fall. Mountain cliffs and roadcuts fall.

The shock waves move up the peninsula. Reaching Los Gatos, they give a wrenching spin to houses that cost seven hundred and fifty thousand dollars and have no earthquake insurance. A man is at work in a bicycle shop. In words that *Time* will print in twenty-four-point type, he will refer to the earthquake as "my best near-death experience." (For a number of unpublished fragments here, I am indebted to editors at Time Warner, who shared with me a boxful of their correspondents' files.)

Thirteen seconds north of the epicenter is Los Altos, where Harriet and David Schnur live. They grew up in New York City and have the familiar sense that an IRT train is passing under their home. It is a "million-dollar Cape Cod," and glass is breaking in every room. This is scarcely their first earthquake.

David: "Why is it taking so long?"

Harriet: "This could be the last one. Thank God we went to *shul* during the holidays."

The piano moves. Jars filled with beans shatter.

Wine pours from breaking bottles. A grandfather clock, falling—its hands stopping at 5:04—lands on a metronome, which begins to tick.

The shock reaches Stanford University, and sixty buildings receive a hundred and sixty million dollars' worth of damage. The university does not have earthquake insurance.

The waves move on to San Mateo, where a woman in a sixteenth-floor apartment has poured a cup of coffee and sat down to watch the third game of the World Series. When the shock arrives, the apartment is suddenly like an airplane in a wind shear. The jolt whips her head to one side. A lamp crashes. Books fall. Doors open. Dishes fall. Separately, the coffee and the cup fly across the room.

People are dead in Santa Cruz, Watsonville has rubble on the ground, and San Francisco has yet to feel anything. The waves approach the city directly from the hypocenter and indirectly via the Moho. Waves that begin this deep touch the Moho at so slight an angle that they carom upward, a phenomenon known as critical reflection. As the shaking begins in San Francisco, it is twice as strong as would generally be expected for an earthquake of this magnitude at that distance from the epicenter.

Two men are on a motor scooter on Sixteenth Street. The driver, glancing over his shoulder, says, "Michael, stop bouncing." A woman walking on Bush Street sees a Cadillac undulating like a water bed. She thinks, What are those people *doing* in there? Then the windows fall out of a nearby café. The sidewalks are moving. Chimneys fall in Haight-Ashbury, landing on cars. In Asbury Heights, a man is watering his patch of grass. He suddenly feels faint, his knees weaken, and his front lawn flutters like water under wind. Inside, his wife is seated at her seven-foot grand. The piano levitates, comes right up off the floor as she plays. She is thinking, I'm good but not this good. A blimp is in the air above. The pilot feels vibration. He feels four distinct bumps.

In Golden Gate Park, high-school girls are practicing field hockey. Their coach sees the playing field move, sees "huge trees . . . bending like windshield wipers." She thinks, This is the end, I'm about to fall into the earth, this is the way to go. Her players freeze in place. They are silent. They just look at one another.

In the zoo, the spider monkeys begin to scream. The birdhouse is full of midair collisions. The snow leopards, lazy in the sun with the ground shaking, are evidently unimpressed. In any case, their muscles don't move. Pachy, the approximately calico cat who lives inside the elephant house, is outside the elephant house. She refused to enter the building all day yesterday and all day today. When someone carried her inside to try to feed her, she ran outside, hungry.

At Chez Panisse, in Berkeley, cupboard doors open and a chef's personal collection of pickles and preserves crashes. The restaurant, renowned and booked solid, will be half full this evening. Those who come will order exceptionally expensive wine. Meanwhile, early patrons at a restaurant in Oakland suddenly feel as if they were in the dining car of a train that has lurched left. When it is over, they will all get up and shake hands.

In the San Francisco Tennis Club, balls are flying without being hit. Players are falling down. The ceilings and the walls seem to be flowing. Nearby, at Sixth and Bluxome, the walls of a warehouse are falling. Bricks crush a car and decapitate the driver. Four others are killed in this avalanche as well.

In the hundred miles of the San Andreas Fault closest to San Francisco, no energy has been released. The accumulated strain is unrelieved. The U.S. Geological Survey will continue to expect within thirty years an earthquake in San Francisco as much as fifty times as powerful. In the Survey's offices in Menlo Park, a seismologist will say, "This was not a big earthquake, but we hope it's the biggest we deal with in our careers." The Pacific Stock Exchange, too vital to suffer as much as a single day off, will trade by candlelight all day tomorrow.

Passengers on a rapid-transit train in a tube under the bay feel as if they had left the rails and were running over rocks. The Interstate 80 tunnel through Yerba Buena Island moves like a slightly writhing hose. Linda Lamb, in a sailboat below the Bay Bridge, feels as if something had grabbed her keel. Cars on the bridge are sliding. The entire superstructure is moving, first to the west about a foot, and then back east, bending the steel, sending large concentric ripples out from the towers, and shearing through bolts thicker than cucumbers. This is the moment in which a five-hundred-ton road section at one tower comes loose and hinges downward, causing one fatality, and breaking open the lower deck, so that space gapes to the bay. Heading toward Oakland on the lower deck, an Alameda County Transit driver thinks that all his tires have blown, fights the careening bus for control, and stops eight feet from a plunge to the water. Smashed cars vibrate on the edge but do not fall. Simultaneously, the Golden Gate Bridge is undulating, fluctuating, oscillating, pendulating. Daniel Mohn—in his car heading north, commuting home—is halfway across. From the first tremor, he knows what is happening, and his response to his situation is the exact opposite of panic. He feels very lucky. He thinks, as he has often thought before, If I had the choice, this is where I would be. Reporters will seek him later, and he will tell them, "We never close down." He is the current chief engineer of the Golden Gate Bridge.

Peggy Iacovini, having crossed the bridge, is a minute ahead of the chief engineer and a few seconds into the Marin Headlands. In her fluent Anglo-Calif she will tell the reporters, "My car jumped over like half a lane. It felt like my tire blew out. Everybody opened their car doors or stuck their heads out their windows to see if it was their tires. There were also a couple of girls holding their chests going oh my God. All the things on the freeway were just blowing up and stuff. It was like when you light dynamite—you know, on the stick—it just goes down and then it blows up. The communication wires were just sparking. I mean my heart was beating. I was like oh my God. But I had no idea of the extent that it had done."

At Candlestick Park, the poles at the ends of the

foul lines throb like fishing rods. The overhead lights are swaying. The upper deck is in sickening motion. The crowd stands as one. Some people are screaming. Steel bolts fall. Chunks of concrete fall. A chunk weighing fifty pounds lands in a seat that a fan just left to get a hot dog. Of sixty thousand people amassed for the World Series, not one will die. Candlestick is anchored in radiolarian chert.

The tall buildings downtown rise out of landfill but are deeply founded in bedrock, and, with their shear walls and moment frames and steel-and-rubber isolation bearings, they sway, shiver, sway again, but do not fall. A woman forty-six floors up feels as if she were swinging through space. A woman twenty-nine floors up, in deafening sound, gets under her desk in fetal position and thinks of the running feet of elephants. Cabinets, vases, computers, and law books are flying. Pictures drop. Pipes bend. Nearly five minutes after five. Elevators full of people are banging in their shafts.

On the high floors of the Hyatt, guests sliding on their bellies think of it as surfing.

A quick-thinking clerk in Saks herds a customer into the safety of a doorjamb and has her sign the sales slip there.

Room service has just brought shrimp, oysters, and a bucket of champagne to Cybill Shepherd, on the seventh floor of the Campton Place Hotel. Foot of Nob Hill. Solid Franciscan sandstone. Earthquakes are not unknown to Shepherd. At her home in Los Angeles, pictures are framed under Plexiglas, windowpanes are safety glass, and the water heater is bolted to a wall. Beside every bed are a flashlight, a radio, and a hard hat. Now, on Nob Hill, Shepherd and company decide to eat the oysters and the shrimp before fleeing, but to leave the champagne. There was a phone message earlier, from her astrologer. Please call. Shepherd didn't call. Now she is wondering what the astrologer had in mind.

A stairway collapses between the tenth and eleventh floors of an office building in Oakland. Three people are trapped. When they discover that there is no way to shout for help, one of them will dial her daughter in Fairfax County, Virginia. The daughter will dial 911. Fairfax County Police will teletype the Oakland police, who will climb the building, knock down a wall, and make the rescue.

Meanwhile, at sea off Point Reyes, the U.S. Naval Ship Walter S. Diehl is shaking so violently that the officers think they are running aground. Near Monterey, the Moss Landing Marine Laboratory has been destroyed. A sea cliff has fallen in Big Sur—eighty-one miles south of the epicenter. In another minute, clothes in closets will be swinging on their hangers in Reno. Soon thereafter, water will form confused ripples in San Fernando Valley swimming pools. The skyscrapers of Los Angeles will sway.

After the earthquake on the Hayward Fault in 1868, geologists clearly saw that dangers varied with the geologic map, and they wrote in a State Earthquake Investigation Commission Report, "The portion of the city which suffered most was . . . on made ground." In one minute in 1906, made ground in San Francisco

sank as much as three feet. Where landfill touched natural terrain, cable-car rails bent down. Maps printed
and distributed well before 1989—stippled and cross-
hatched where geologists saw the greatest violence to
come—singled out not only the Nimitz Freeway in
Oakland but also, in San Francisco, the Marina district,
the Embarcadero, and the Laocoönic freeways near
Second and Stillman. Generally speaking, shaking de-
clines with distance from the hypocenter, but where
landfill lies on loose sediment the shaking can amplify,
as if it were an explosion set off from afar with a
plunger and a wire. If a lot of water is present in the
sediment and the fill, they can be changed in an instant
into gray quicksand—the effect known as liquefaction.
Compared with what happens on bedrock, the damage
can be something like a hundredfold, as it was on the
lakefill of Mexico City in 1985, even though the hypocenter was far to the west, under the Pacific shore.

In a plane that has just landed at San Francisco International Airport, passengers standing up to remove
luggage from the overhead racks have the luggage removed for them by the earthquake. Ceilings fall in the
control tower, and windows break. The airport is on
landfill, as is Oakland International, across the bay.
Sand boils break out all over both airfields. In downtown San Francisco, big cracks appear in the elevated
I-280, the Embarcadero Freeway, and U.S. 101, where
they rest on bayfill and on filled-in tidal creek and
filled-in riparian bog. They do not collapse. Across the
bay, but west of the natural shoreline, the Cypress

section of the Nimitz Freeway—the double—decked I-880—is vibrating at the same frequency as the landfill mud it sits on. This coincidence produces a shaking amplification near eight hundred per cent. Concrete support columns begin to fail. Reinforcing rods an inch and a half thick spring out of them like wires. The highway is not of recent construction. At the tops of the columns, where they meet the upper deck, the joints have inadequate shear reinforcement. By a large margin, they would not meet present codes. This is well known to state engineers, who have blueprinted the reinforcements, but the work has not been done, for lack of funds.

The under road is northbound, and so is disaster. One after the last, the slabs of the upper roadway are falling. Each weighs six hundred tons. Reinforcing rods connect them, and seem to be helping to pull the highway down. Some drivers on the under road, seeing or sensing what is happening behind them, stop, set their emergency brakes, leave their cars, run toward daylight, and are killed by other cars. Some drivers apparently decide that the very columns that are about to give way are possible locations of safety, like doorjambs. They pull over, hover by the columns, and are crushed. A bank customer-service representative whose 1968 Mustang has just come out of a repair shop feels the jolting roadway and decides that the shop has done a terrible job, that her power steering is about to fail, and that she had better get off this high-speed road as fast as she can. A ramp presents itself. She swerves onto it and off the freeway. She hears a huge sound. In her rearview mirror she sees the upper roadway crash flat upon the lower.

As the immense slabs fall, people in cars below hold up their hands to try to stop them. A man eating peanuts in his white pickup feels what he thinks are two flat tires. A moment later, his pickup is two feet high. Somehow, he survives. In an airport shuttle, everyone dies. A man in another car guns his engine, keeps his foot to the floor, and races the slabs that are successively falling behind him. His wife is yelling, "Get out of here! Get out of here!" Miraculously, he gets out of here. Many race the slabs, but few escape. Through twenty-two hundred yards the slabs fall. They began falling where the highway leaves natural sediments and goes onto a bed of landfill. They stop where the highway leaves landfill and returns to natural sediments.

Five minutes after five, and San Francisco's Red Cross Volunteer Disaster Services Committee is in the middle of a disaster-preparedness meeting. The Red Cross Building is shivering. The committee has reconvened underneath its table.

In yards and parks in the Marina, sand boils are spitting muds from orifices that resemble the bell rims of bugles. In architectural terminology, the Marina at street level is full of soft stories. A soft story has at least one open wall and is not well supported. Numerous ground floors in the Marina are garages. As buildings collapse upon themselves, the soft stories vanish. In a fourth-floor apartment, a woman in her kitchen has been cooking Rice-A-Roni. She has put on long johns and a sweatshirt and turned on the television to watch the World Series. As the building shakes, she moves with experience into a doorway and grips the jamb. Nevertheless, the vibrations are so intense that she is thrown to the floor. When the shaking stops, she will notice a man's legs, standing upright, outside her fourth-story window, as if he were floating in air. She will think that she is hallucinating. But the three floors below her no longer exist, and the collapsing building has carried her apartment to the sidewalk. Aqueducts are breaking, and water pressure is falling. Flames from broken gas mains will rise two hundred feet. As in 1906, water to fight fires will be scarce. There are numbers of deaths in the Marina, including a man and woman later found hand in hand. A man feels the ground move under his bicycle. When he returns from his ride, he will find his wife severely injured and his infant son dead. An apartment building at Fillmore and Bay has pitched forward onto the street. Beds inside the building are standing on end.

The Marina in 1906 was a salt lagoon. After the Panama Canal opened, in 1914, San Francisco planned its Panama-Pacific International Exposition for the following year, not only to demonstrate that the city had recovered from the great earthquake to end all earthquakes but also to show itself off as a golden destination for shipping. The site chosen for the Exposition was the lagoon. To fill it up, fine sands were hydraulically pumped into it and mixed with miscellaneous debris, creating the hundred and sixty-five dry acres that flourished under the Exposition and are now the Marina.

Nearly a minute has passed since the rock slipped at the hypocenter. In San Francisco, the tremors this time will last fifteen seconds. As the ground violently shakes and the sand boils of the Marina discharge material from the liquefying depths, the things they spit up include tarpaper and bits of redwood—the charred remains of houses from the earthquake of 1906.

An earthquake.

A small flex of mobility in a planetary shell so mobile that nothing on it resembles itself as it was some years before, when nothing on it resembled itself as it was some years before that, when nothing on it . . .

Not long ago, at Mussel Rock, a man named Araullo was fishing. He had a long pole that looked European. He seemed not so much to be casting his lure as sweeping it through the sea. His home was near the top of the cliff. He pointed proudly. The one nearest the view.

He had come down the trail and jumped over water to a wide, flat boulder. The seismic crack that came down the cliff ran into the water and under the boulder. He was fishing the San Andreas Fault, and he was having no luck.

I asked him, "What are you after?"

He said, "Sea perch. I also get salmon and striped

bass here. Now I don't know where they are. Someday, they come."

He said that he felt very fortunate to have a house so close to the fish and the ocean, to have been able to afford it. He had bought it six months before. In this particular location, real estate was cheap. He had bought the house for a hundred and seventy thousand. I could barely hear him over the sound of the waves.

"If it going to go down, it going to go down," he shouted, and he flailed the green sea. "You never know what going to happen. Only God knows. Hey, we got the whole view of the ocean. We got the Mussel Rock. What else we need for? This is life. If it go down, we go down with it."

The cormorants were present, and the pelicans. The big fishing boulder was echeloned with shears. From somewhere near Araullo's house, a hang glider had left the jumpy earth and now hovered safely above us.

Araullo ignored the hang glider and kept on swinging his pole.

"I don't know where they are," he said again. "But someday they come. They always come."